THAILAND'S CORAL REEFS

NATURE UNDER THREAT

THAILAND'S CORAL REEFS

NATURE UNDER THREAT

Collin Piprell

Ashley J. Boyd

White Lotus
Bangkok Cheney

Acknowledgments

We would like to express our gratitude to the following people for agreeing to read the manuscript: Dr Hansa Chansang, Senior Marine Biologist at the Phuket Marine Biological Center; Dr Donald MacIntyre, of the Institute of Agriculture, Sterling University, Scotland; and Dr Suraphol Sudara, of the Department of Marine Science, Chulalongkorn University, Bangkok. Khun Sompong Asawajittanont of the Coastal Resources Management Project, Office of the National Environment Board, provided much-appreciated data concerning the distribution and condition of Thai reefs.

Special thanks are due Dr Hansa Chansang for her particularly careful comments. She also helped in the identification of species pictured in this book, together with her colleagues at the Phuket Marine Biological Center Ukrit Satapoomin (fish and octocorals), Niphon Pongsuwan (hard corals and octocorals), and Somchai Bussarit (other phyla). The Phuket Marine Biological Center was also kind enough to provide common Thai names for some of the marine organisms in these photographs.

The book has also benefited from Manfred Winkler's careful drawings and maps.

White Lotus Co., Ltd
G.P.O. Box 1141
Bangkok 10501
Thailand

Printed in Thailand

Typeset by COMSET Limited Partnership

ISBN 974-8496-42-2 pbk. White Lotus Co., Ltd., Bangkok
ISBN 1-879155-55-9 pbk. White Lotus Co., Ltd., Cheney

CONTENTS

ILLUSTRATIONS

FIGURES

MAPS

INTRODUCTION: SAVING THAILAND'S REEFS

The coral reef is one of nature's richest field laboratories, a great genetic archive matched only by that of the tropical rain forest. It holds the potential for untold numbers of insights into the origins and processes of life, into the questions of who we are and why and what we might be—contributions to human knowledge which have a value not to be measured simply in terms of money.

Yet the incredible richness and beauty of a healthy coral reef may be something our grandchildren experience only in films and books. A wealth of species, many of them not yet identified, will soon have disappeared forever from the seas of this planet. Whole ecosystems are likely to have become extinct. We possess the means to avert this tragic loss; but have we the will?

Denuded hillsides stippled with treestumps and slashed by landslides have become daily images in the newsmedia, given growing public alarm over Thailand's fast-dwindling forests. Scenes of equivalent blight underwater have received less attention. This book is intended both as an introduction to the natural history of the coral reef for some readers and as a new look at this underwater world for others. At the same time a plea is made for the preservation of this precious resource, a resource which is threatened with annihilation.

Reefs have survived natural catastrophes for many millions of years. Vast areas have been wiped out by such contingencies as volcanic action and violent storms, as well as by temperature changes and fluctuating sea-levels associated with the advance and retreat of the ice caps. Millions of species of marine organisms have become extinct over the ages. So what's different now? one might ask. What novel factors have entered the equation such that one can suggest a special crisis exists today? Basically, the problem is humankind and its technology.

In the first place, human populations have exploded to the point our predations are exerting severe pressure on other species and the habitats we often share with them. Secondly, our technology is spewing billions of tons of wastes into the environment, poisoning the forests and seas, the very cradles of life on Earth.

But why save the reefs? What good are they, anyway? This might well be the next question, especially from one who has never even seen a reef. Let us explore more specifically, then, just what sort of resource a coral reef represents from the human perspective.

First of all, of course, the reef supports a variety of marine organisms which are valuable to local economies. The US-ASEAN Coastal Resources Management Project has concluded that tropical coral reefs are the most productive of the various coastal ecosystems, contributing 10-30 percent of the total fisheries catch annually. Further, it could be suggested that a cultural value is at risk, one greater perhaps than the commercial one. About 2,000 people on the southern island of Phuket alone make at least part of their living as small-scale fishermen on the reefs. Already they report having to travel farther and farther offshore to more and more remote reefs. Destroy the reef, and the traditional livelihood and lifestyle of the local fishermen is ended.

But reefs have still wider significance than that to the fishery industry: large-scale commercial fishing is also affected. Many oceanic species of fish depend for some part of their life cycle on the reef, the inshore habitats functioning, for instance, as 'nurseries' for their fry.

Fringing reefs, moreover, are a source of sand to replenish beaches at the same time they protect harbors, beaches and islands from wave erosion, helping preserve valuable real estate. Reefs are themselves an important recreational resource, with an estimated 75,000 tourists annually snorkeling or scuba diving in the Andaman Sea alone, and with that number growing year by year.

*The coral reef is one of the two richest habitats on Earth, an exuberant evolutionary experiment, a vast bank of genetic material, a storehouse of potential medical and other scientific discoveries. (***Opposite page:** *Koh Payu, Similans, in 32 meters.* **Above:** *Christmas Reef, Koh Bangu, in 15 meters.) Dynamite fishing or extensive anchor damage can result equally in scenes of desolation on what once were thriving coral reefs. (***Below:** *Koh Hu Yong, Similans, in 6 meters.)*

In the longer term, many marine organisms which depend on this specialized environment could well provide such benefits as new medicines. Already the vast potential pharmacopoeia represented by sponges, algae, and other reef creatures has yielded promising substances in the treatment of cancer and AIDS, for just two examples. The next species or genus to become extinct might have held a cure for cancer — only by then it will have been too late, and we will never know.

Several good reasons for preserving our reefs have been proposed without as yet leaving the realm of commercial arguments. And these considerations aside, this generation should not want to be responsible for denying those to come the sense of exquisite beauty and wonder the reef inspires at first hand. No photographs or words can begin to do justice to this experience of the amazing mystery and diversity of life.

One purpose of this book, then, is to acquaint the reader with a resource which is usually out of sight, and consequently all too often out of mind. It is not meant to compete with the many excellent textbooks on marine biology and coral reefs which are already available. Presupposing a literate reader, though not necessarily one with a scientific background, Part One describes in simple terms how the reefs found in Thai waters first arose, how they have acquired their characteristic shape and distribution, and how the various resident marine organisms are related within a complex ecosystem. We see how the decimation of one species can have drastic and unforeseen consequences in other areas of the ecosystem. The reader is also referred to an appendix wherein eleven groups of marine animal commonly found in Thai waters (and generally accessible to snorkelers and scuba divers alike) are illustrated and described in broad outline.

Parts Two and Three look at the various sources of reef damage before going on to investigate past and current conservation efforts in Thailand.

Part Four offers an assessment of these efforts, concluding with an argument that any optimistic prognosis presupposes widespread and profound attitudinal changes in the public at large. Given the rate at which we are destroying our coral reefs on one hand, and the usual pace of cultural change on the other, it is difficult not to remain pessimistic regarding chances for the survival of a large proportion of these resources.

PART ONE

A NATURAL HISTORY OF THE THAI REEF

The hard coral Symphyllia sp., *a species of 'brain coral' (Hin Pae, Chumphon, in 12 meters).*

A variety of acroporas are common in Thai waters. This coral was found off Koh Sak, Pattaya, in 9 meters.

Other hard corals, Fungia sp., *are commonly referred to as mushroom corals.* **Left:** *This one was spotted off Koh Nok, Samae San, in 15 meters.* **Right:** *Opal bubble corals such as* Plerogyra sinuosa *are sometimes referred to as 'grape corals'. The polyps are extended only at night. The vesicles — the 'bubbles' seen in this photo — are deployed during the day and hide the skeleton.* Plerogyra *are among the few corals well enough armed with nematocysts to sting a human. This specimen was spotted off Hin Pae, Chumphon, in 17 meters.*

Above: *The polyps of the gorgeous* Tubastrea sp., *or 'daisy coral', emerge both at night and, at least on the walls of submarine caves and in shady areas, during the day. (Koh Larn Kai, Chumphon, in 15 meters.)* **Right:** *The antipatharians, commonly referred to as 'black corals', take either bushy or whip-like forms. Like the other antipatharians, this sea whip* (Stichopathes sp.) *can thrive at depths where, cut off from sufficient light to support their photosynthetic algal partners, the stony corals cannot. (Koh Ba Ngu, Similans, in 15 meters.)* **Below:** *Damselfish hover around a well-developed table coral off Koh Hu Yong, Similans, in about 11 meters.*

A NATURAL HISTORY OF THE THAI REEF

Imagine extraterrestrial visitors standing on the Moon and looking up at the planet we call Earth. Only two features visible to the naked eye would reveal the presence of life. One is a human artifact—the Great Wall of China. The other is the Great Barrier Reef, off the eastern coast of Australia. Were they tempted to investigate more closely, those alien visitors would surely find much more to wonder at.

To a similar extent, perhaps, the coral reef presents an exotically beautiful and alien world to us here on our own planet, if only we look.

Most of us, of course, have never had the opportunity to dive on coral reefs; we know little about their history and nature. Neither do most of us realize that the tropical reef and its hundreds of thousands of denizens are threatened with widespread destruction by the only other creatures on Earth capable of building on such a grand scale.

One's first impression of a coral reef is that it is a riot—that this is evolution gone wild, a bedlam of different species and shapes and colors and patterns. Nevertheless, although many mysteries remain, marine scientists know enough to say that there is sublime organization behind the apparent chaos.

Without a deliberate plan, with no central nervous system to coordinate activities, the needs of the various reef organisms, constrained by natural forces in the environment, have somehow shaped a largely self-contained community within which a great number of species play their own vital roles. The success of the coral reef as a whole depends upon the success of the wide range of different organisms which in turn are supported by the reef habitat. The entire reef, then, can be interpreted as a single creature—just as we normally think of the hard coral as one animal, no matter it is in reality a colony of mutually dependent organisms. Given the astonishingly complex relationships and activities which allow the reef to build and maintain its structure, that substratum upon which its vital components may produce food, handle wastes, and recycle essential nutrients, the reef as a whole may be said to have a 'metabolism' of its own.

Any discussion of the coral reef must begin with an account of the hard corals themselves. This remarkable instance of a mutualistic relationship between species, the one a plant and the other an animal, is, after all, perhaps the most important element in the formation of reefs. Without this symbiotic relationship, the corals might not even have gained a foothold on the environmental space which is their chosen niche, much less thrived there for hundreds of millions of years.

Soft corals, black corals, and hard corals are typically associated with the reef, but only the hard corals are termed 'hermatypic', or reef builders. These corals are, first of all, of the phylum Co*elenterata* (or *Cnidaria*)— that is to say, in the Greek, 'with a hollow gut'. They are at the same time of the class A*nthozoa*, or 'flower animals'. Finally, they are of the order *Scleractinia*, or 'hard'. Hermatypic corals, in short, are hollow-gutted flower-like animals with a hard exoskeleton into which they can retreat when threatened (see Figures 1 and 2). Over 200 species, belonging to 75 genera, have been recorded in the Phuket area alone; 60 species have so far been noted in the Gulf of Thailand.

The hard corals are chiefly distinguished from other coelenterates by two features: most of them play host to symbiotic unicellular algae called zooxanthellae, and they produce skeletons of calcium carbonate which over time can form large accretions—the stony reef substratum itself. Hard corals are closely related to the sea anemones (and, more distantly, to the jellyfish and hydroids). Indeed, they resemble nothing so much as colonial anemones supported and protected by the calcareous skeletons which they secrete. Though some corals are solitary,

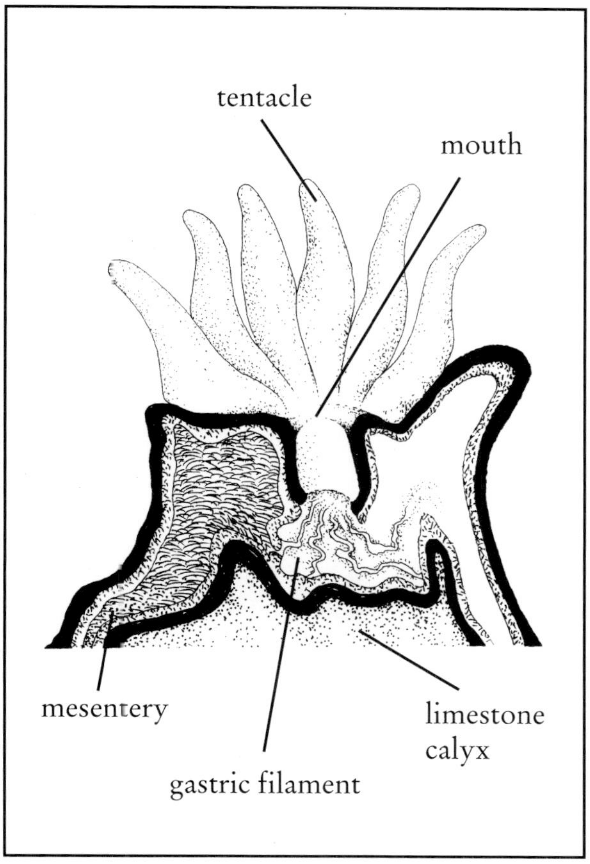

Figure 1: Coral polyp

with polyps up to 30 cm in diameter, most are colonial, with each colony comprising thousands of polyps, each polyp averaging 2-10 mm in diameter, depending on the species.

The hard corals display hexameral symmetry; their tentacles come in multiples of six and these are arrayed in one or more circles around a central mouth. Armed with stinging cells, or 'nematocysts'—spring-loaded capsules containing harpoon-shaped barbs on filaments—the tentacles are employed first to trap, incapacitate, and finally convey zooplankton to the mouth. The mouth opens on an esophagus which leads to a gastrovascular cavity, or stomach. (Both 'mouth' and 'stomach', we will see later, in fact have multiple functions.) The polyps are attached to the hard mass of the reef by basal disks; and they are protected by cup-like cells of calcium carbonate which they secrete and which over time are deposited one atop the other to build up reefs which may be hundreds of meters thick. The living colony itself is only a thin skin of polyps growing on the matrix of the dead substratum.

When the polyps divide, an intimate identity is maintained—the parent polyp stays joined to the new individual by an extension of its stomach cavity. Through this network of connecting stomach extensions, food is shared out all over the colony, with each polyp both contributing to and sharing in the general welfare.

Tiny as the individual coral organisms are, and although they are normally fixed to one spot for most of their life cycles (until 250 years ago, biologists had assumed they were plants), they are key builders of features on the scale of Australia's Great Barrier Reef, a massive formation 2000 km long and 160 km across at its widest. These reefs are largely composed of the calcium carbonate skeletons which the corals produce, and for this production they depend on a symbiotic relationship with zooxanthellae, a unicellular algae which inhabits the tissues of the hermatypic corals. As we shall see, it is this relationship which has allowed the hard corals to claim an ecological niche which would be otherwise uninhabitable by any known organism.

Figure 2: Limestone calyx

Above: *Among the most colorful animals on the reef,* Dendronephthya *sp. lack the external skeletons of their stony cousins the hexacorals. Soft coral colonies are instead given a certain amount of rigidity by sclerites, hard crystalline structures imbedded in their tissues. (Similan Islands in 15 meters.)* **Left:** *Like the soft corals, sea fans such as this* Melithaea sp. *are octocorals—they display eight-sided, rather than the six-sided symmetry of the hexacorals. In addition to the sclerites found in the soft corals, gorgonian fan corals and sea whips also have a second skeletal structure, a hard internal structure of gorgonian. (Koh Similan, 'Island No.8', in 24 meters.)* **Below:** *A close-up of another species of gorgonian shows the polyps fully extended in search of food. Immediately something comes in contact with the colony, the polyps retract. This specimen is growing off Koh Ngam Yai, Chumphon, in 14 meters.*

The coral reef supports a staggering number of little worlds, many of them including very different species living together for mutual benefit. Here the diver (Homo sapiens) *examines an anemonefish* (Amphiprion akallopisos) *which goggles right back, confident of the protection afforded by the stinging tentacles of its host sea anemone* (Radianthus sp.). *What advantage does the anemone get out of the partnership? Perhaps the fish provides cleaning services. (Koh Similan, Campbell's Bay, in 9 meters.)*

Chapter I

Oases in a Marine Desert

A symbiotic community at the cellular level. The symbiosis between hermatypic corals and zooxanthellae makes possible animal life in waters which might otherwise be barren. Because the coral can first of all produce its own food where carbon dioxide and sunlight are abundant, and because the coral ecosystem, as we shall see, is self-contained and capable of recycling scarce nutrients, the reef is an oasis of biological activity. More than that it is— with the single exception of the tropical rain forest—biologically the richest habitat on Earth, supporting hundreds of thousands of species.

A hermatypic coral polyp is made up of two layers of cells, with the inner layer hosting the algae. How this association comes about in the first place has not yet been established, but it has been discovered the relationship may be terminated under conditions of stress, such as when the coral is starved for food or at the rapidly growing tips of a staghorn coral. (Why this should be so is not understood. As of this writing, a recent global warming of the seas is causing this sort of 'bleaching' of hard corals around the world, including in Thai waters.) Outside their host, zooxanthellae metamorphose; they change from the spherical shape they typically assume within the coral to that of a dinoflagellate, the typical free-swimming algal plankton propelled by twin whip-like appendages.

Whereas the zooxanthella can under certain circumstances revert to a free stage and exist

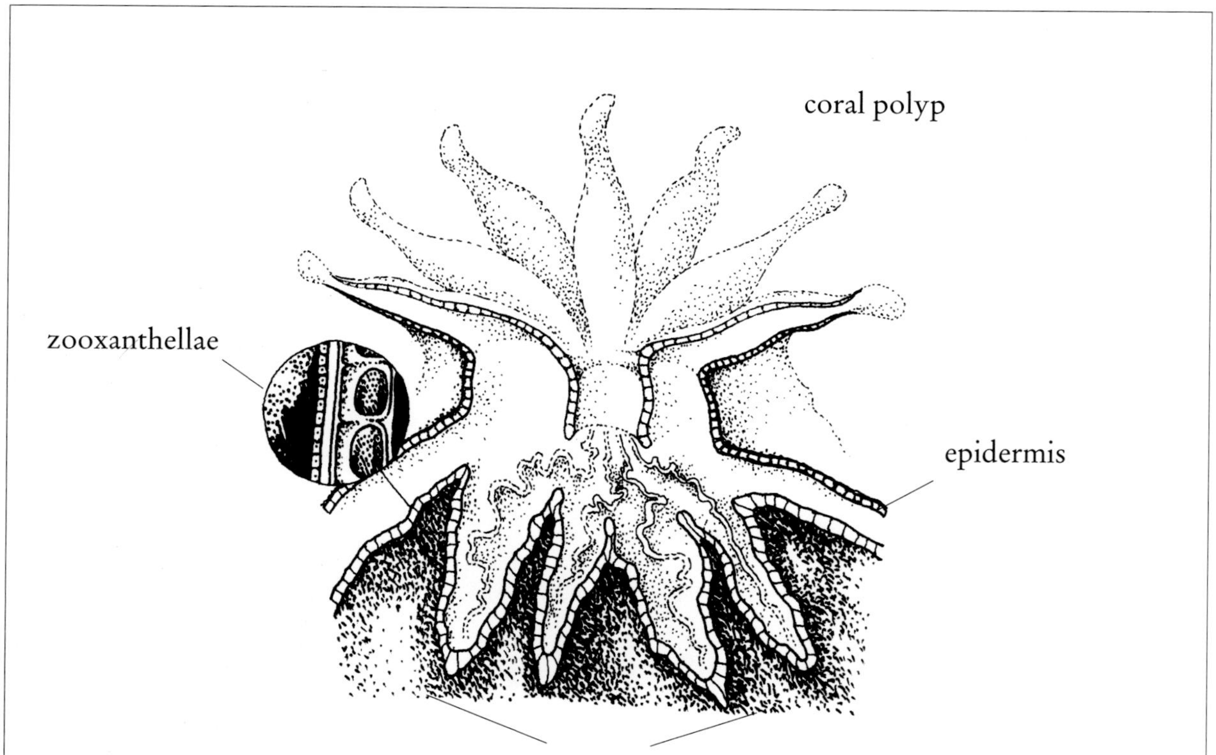

Figure 3: A fundamental plant-animal collaboration

independently of its host, the reef as a whole is dependent on its relationship with the algae. The zooxanthella contributes to the success of the coral in at least four different ways: as a supplementary source of nutrients; as a means of recycling wastes; as an additional source of energy; and as a contributor to calcification in the colony. Let us look at each of these factors in turn.

Zooxanthellae are photosynthetically active: they convert sunlight and carbon dioxide into chemical energy—into sugars, amino acids, and fats. The gold-brown pigments which color these algae are ideal for the absorption of the blue wave-lengths of sunlight which best penetrate water. So long as they have adequate carbon dioxide and sun, the algae can go on producing food and oxygen forever. Experiments have indicated that as much as four-fifths of the algae's production is put at the disposal of their hosts (the zooxanthella cell leaks nutrients when stimulated by digestive enzymes secreted by the coral polyps).

Zooxanthellae are also involved in the recycling of wastes. The coral might suffocate in its own ammonia waste products, without the activity of its symbiont. At the same time, the algae absorbs acetate, the end product of food oxidation by the host, and uses it in the creation of fatty acids essential for its own health.

Some of these fatty acids are passed along to the coral host, which in turn converts them into a kind of lipid, a waxy substance similar to the ambergris found in the cranial cavities of sperm whales and which is readily converted to energy. In fact, it has been estimated that these lipids can constitute as much as one-third of the dry weight of corals. Until recently, it was a mystery why the coral would need so much in excess of its apparent energy requirements. Scientists have since discovered, however, that the larvae periodically dispersed by some corals to establish new colonies are loaded with this 'fuel'.

Finally, the algae contribute to effective calcification in the coral colony. They remove phosphate from the wastes of their host, and dissolved phosphate has been shown to inhibit calcium carbonate precipitation. It is also possible, though not yet proven, that zooxanthellae contribute proteins which help provide an organic matrix that must be deposited before the calcium carbonate which is secreted by the coral's basal disk may crystallize successfully.

It is too soon to say whether any one of these advantages is primary, or precisely why evolution has selected for this mutualistic relationship. Zooxanthellae are also found in symbiotic relationships with some anemones (and other organisms such as giant clams and certain gastropods). Given that sea anemones are closely related to the hard corals, and given that anemones do not secrete skeletons, it might appear that nutrition and energy are the major determinants. Provisionally, however, marine biologists conclude that zooxanthellae probably lend different primary advantages in different relationships, depending on the nature and habitat of the host organism. With the hard corals, of course, their contribution to calcification is relatively important. Where other food sources (i.e. zooplankton and dissolved organic matter) are scarce, however, the provision of supplementary nutrients and energy assumes a more vital role.

What benefits does the algae derive from this relationship? First of all, we may suppose, the zooxanthellae enjoy shelter from their predators. Beyond that adaptive advantage, however, they gain access through the metabolic wastes produced by the coral host to essential nutrients which might not be available in the open water. Phosphorus is one such nutrient. So is acetate, as we have seen, which is used in the production of essential fatty acids. Nitrogen is produced, furthermore, from the ammonia wastes which would otherwise poison the host. Just as the corals which have adapted to these oceanic oases could not survive in the environment they favor without their algal partners, so the algae would not continue to thrive in those otherwise barren waters without the advantages offered by their hosts.

The larger community: the entire reef considered as a single organism. An oceanic reef (an atoll or barrier reef) is virtually self-sufficient, generating most of its own food. Its main needs are simple: sunlight, as the source of energy for photosynthesis, together with carbon dioxide and calcium, both of which are plentiful in seawater. It also needs phosphorus and nitrogen, however, which are readily available though not in forms in which they can be used. Some process

Above: *This xanthid coral crab* (Trapezia cymodoce) *lives in an interesting relationship of mutual advantage with the shrub-like coral* Pocillopora damicornis *and a little community of different species which inhabit the same coral thicket at once. In exchange for the protection afforded by the close branches of the host, both pistol shrimp and crabs help defend the coral against the predations of crown-of-thorns starfish* (Acanthaster planci). *Should one of these scourges of the hard corals attempt to attack their host, the shrimps* (Alpheus luttini) *snap at the intruder till it withdraws, while the crabs grab the feet or spines of the starfish and jerk them up and down till the intruder moves off. (Hin Pae, Chumphon, in 15 meters.)* **Below:** *Some fish, immune to stinging cells which pack enough poison to discourage most predators, find refuge among the tentacles of drifting jellyfish. Unless the fish act as bait for other fish which are not immune, it is not clear what advantage accrues to the jellyfish in this relationship. (Koh Rin, Sattaheep, in 12 meters.)*

Above: *Cleaner wrasse such as this* Labroides dimidiatus *have evolved another kind of immunity. Advertising their services partly through color and pattern, partly through distinctive ritual movement, they persuade larger fish, in exchange for the removal of parasites and dead tissue, to forego the natural temptation to make a snack of the wrasse. (Koh Ba Ngu, Similans, in 12 meters.)* **Below:** *Banded cleaner shrimp* (Stenopus hispidus) *typically work the graveyard shift at cleaning stations on the reef, taking over after the cleaner wrasse have retired for the night. Again, a combination of color, pattern, and distinctive movement protects them from their own customers. (Koh Ba Ngu, Similans, at night in 12 meters.)* **Inset:** *'Shrimp gobies'* (Ambleleotris sp.) *share their burrows in the sandy bottom with a type of shrimp of the Alpheidae family. After the shrimp has scooped out the sand, it takes one piece of shell or stone back down, perhaps to shore up the burrow. (Koh Ngam Yai, Chumphon, in 14 meters.)*

is required to turn these elements into the phosphates and nitrates essential to cellular growth. Finally, certain trace elements are needed as well, and these are often very scarce, so the reef ecosystem must conserve the resources it already has. Thus we find three major functional requirements: provision of food, the recycling of scarce essentials, and the building and maintenance of the reef structure.

Beneath all the flash and dazzle of the reef fish, behind all the exotic variety and fascination of the invertebrate species, lie the actual primary producers of the reef, modest and often unremarked. The zooxanthella, hidden within its tiny polyp host, is one such pillar of the community. Another is filamentous algae, the soft fuzz which tends to grow over every available surface on the reef. Beds of sea grass also offer rich grazing to both fish and invertebrates. The blue-green algae are vital contributors as well, though they appear to the uneducated eye as no more than black or reddish stains on rock and coral. They are much more than that, however: they are in essence fertilizer factories, converting nitrogen into nitrates. There is indeed more than meets the eye to this wonderfully integrated and balanced ecosystem.

These various photosynthetic plants, then, convert sunlight and carbon dioxide from seawater into both food and reef substrate, while nitrogen from air and the ammonia in organic waste is turned into nitrates which fertilize the producers themselves.

The symbiotic partnership between hermatypic corals and zooxanthellae establishes the basis for a community which includes other organisms numbering in the hundreds of thousands of distinct species. Again, mutualism obtains at a variety of levels—survival in this environment requires a high degree of collaboration, however unwitting or purely fortuitous.

We have seen how the hard corals and zooxanthellae cooperate to produce nutrients and at the same time contribute to the hard reef substrate. It has also become clear that this mutualistic relationship is only part of a much larger and much more complex ecosystem, one which has so far been only partly described and which is only poorly understood. In fact, it seems that there are hierarchies of interdependence, relationships of dependence within dependences within dependences like Russian dolls. The very basic symbiosis between coral and zooxanthellae in a single coral colony can find itself the substrate for many other communities.

A rather specially charming example is the shrub-like coral *Pocillopora damicornis* and the little community of fishes, crabs, and shrimp which it often hosts, with several species inhabiting the same coral thicket at once. In exchange for the protection afforded by the close branches of the host, the shrimp and crabs help defend the coral against the predations of crown-of-thorns starfish (*Acanthaster planci*). Should one of these beautiful monsters, these banes of the hard corals, attempt to attack this particular happy family, the pistol shrimps (*Alpheus luttini*) snap at it till it withdraws, while the crabs (*Trapezia*) grab the feet or spines of the starfish and jerk them up and down till the intruder moves off. (An interesting note: the crabs also attack newcomer fish, but leave 'residents' alone, the latter individuals signaling with a shivering movement their membership in the community.)

The crown-of-thorns, meanwhile, is actually related to the coral community within a larger relationship, one not so obviously functional for the whole system or so clearly defined, perhaps, but one still necessary for the overall health of the reefs. In some opinion, the crown-of-thorns helps prevent the dominance of any particular corals, thereby contributing to an advantageous variation of species on a reef. The shrub coral, as we have seen, is itself a basic part of a complicated system which maximizes both production and conservation of fuel and stony substrate. (Its resident fish, crabs, shrimp, and predator starfish are ultimately part of the system as well).

From the photosynthetic plants, the primary food producers, right up through the food chain, the survival of the reef depends on a delicate balance of mutual interdependences on the one hand, and competitive relationships on the other. Various organisms feed on the plants. Predators, in turn, feed upon these plant-eaters. Microbes feed on the waste and dead tissues. Each species has its place in the cycle, and it is generally kept in balance, with booms in individual species normally being short-lived, and natural forces of one sort or another contributing to the stability of the ecosystem as a whole.

Chapter II

24 Hours on the Reef

The reef never sleeps. In the course of a 24-hour cycle, it offers a variety of faces to the observer, and we are going to sample these with four dives: one at night, another in the early morning, a third around midday, and one more at dusk. (Of course, other factors from the tides to the seasons will also influence activity within this community, and the account which follows is a vast simplification of the daily routine on a reef.)

Night: Looking down from the boat on the still surface of a tropical lagoon at night, you can see signs of life in the inky black. Sudden serpentine trails of phosphorescence, tiny flashing bits of purple glitter suggest a hidden realm of activity, the mysterious comings and goings of unseen creatures.

Better yet, don diving gear and go down to explore. Turn off your underwater torch and enjoy even more of the bio-luminescent spectacle—pulsating balls of ghostly white fire (jellyfish), abrupt streaks of light (squids or fish), winking points (zooplankton), and other, sometimes strange and unidentifiable patterns of light which might simply be dead fish in the process of being consumed by light-emitting bacteria. Sweep your arm through the water in front of you, and watch the soft wash of phosphorescence which follows. Turn on your torch again to catch the agent of one particularly impressive serpentine trail, and see... nothing. There is no neon snake, no new and wonderful animal. It was merely a fish, already long gone, darting past and firing a path of phosphorescent plankton.

Switching on your electric torch once more, you set about exploring the nocturnal reef in earnest. This is not the same reef you see by day. The first thing you notice is how much more vivid everything is. The colors are dazzling.

But surely, you might suggest, everything should be brighter in *daylight*? In fact that is not true, except at shallow depths in full sunlight. As you descend, the water progressively absorbs longer wavelengths of light, with the reds and yellows going first. You don't have to go very deep before a fish (or a person, for that matter) bleeds green. At depths of 30 meters and more, most things are rendered in shades of blue and black. After dark, on the other hand, your torch restores the scene to its true colors at whatever depth you happen to be. You travel along in a cocoon of light, constantly diverted by the exotic creatures which impinge on your cozy little universe.

Many of these animals are the same ones you see during the day. Often, however, their appearance or their behavior has changed. You may see a big, heavy-bodied moray eel swimming free, for instance. During the day, you will rarely see more than its head poking cautiously out of its hole in the reef. At night, it comes out to hunt. Or sometimes, just as a break from the serious matter of finding nourishment, and much like a businessman stopping for a midday haircut and manicure, the moray may linger at a night 'cleaning station', enticed by the advertising display of the attendant cleaner shrimp as they wave their long, conspicuous antennae. (This is a rather civilized example of mutualism: the fish forego the convenient snack these shrimp might otherwise represent, and instead allow the shrimp to dine on the parasites and bits of dead tissue which adhere to the fish.)

The reef as a whole never sleeps, but many of its denizens do. The parrotfish, for one, can usually be found dead to the world, comfortably esconced up under a ledge in the tent of mucus which it exudes at night, a delicate shroud designed to reduce vulnerability to night-hunting predators (such as the moray) which home in on

Left: *Dawn's early light, and the reef comes alive with fish. The night hunters are still at large, while species that rest at night are also out and about and hungry. At the same time, many larger species which typically feed at dawn and dusk are in evidence. Threatened by fish species which hunt by sight, meanwhile, other species begin to school. (Koh Miang, Similans, in 12 meters.)* **Below:** *Daybreak changes the behavior of many smaller fish, which begin to school in the morning. Schooling reduces the effective threat from fish who hunt by sight, since these predators usually go for individuals, and seem confused by the choice a school offers. An exception is this* Pterois volitans. *The zebrafish merely hangs about resembling a clump of floating weed, waiting for smaller fish to swim within range of its lightning quick, vacuum-cleaner-like attack. (Koh Similan, Campbell's Bay, in 12 meters.)*

Opposite page, top: *Many marine animals behave differently at night and in the day. After dark this parrotfish* (Scarus niger), *for example, secretes a tentlike mucous shroud within which it sleeps, secure from the predations of nocturnal creatures such as moray eels which hunt by scent. (Koh Miang, Similans, in 14 meters.)* **Opposite, bottom:** *This giant hermit crab* (Dardanus megistos) *is just one of many crustacean species which generally come out of hiding to forage only after dark. This one was ambushed by the camera in about 20 meters on the wreck of the* Bremen, *off Sattaheep.*

their prey by scent. (Generally, animals which hunt by sight in the day sleep at night.) The parrotfish sleeps wide-eyed but soundly, so soundly you can gently reach under and lift the fish without waking it. Gently is the operative word, though, for if suddenly awakened it takes off like a torpedo—with enough force, in fact, to shatter the face-plate on your mask.

You see lots of red fish at night, by the light of a torch. Species such as squirrelfish and soldierfish hunt at night because the color red is filtered out at these depths—the fish appear black to their enemies, which is excellent night camouflage. In fact, many marine species display coloration and patterns which are highly conspicuous at one time of day, but which provide excellent protection at another. This indicates that their predators (or, perhaps, their prey) are asleep or elsewhere for some part of the day. (An example of this type of adaptation among land animals is the zebra. The zebra stands out very distinctly on the savannah in the middle of the day, but its predators are resting at this time. In the evening, on the other hand, when the lions and leopards begin to feed, the zebra is almost invisible.)

Other species actually change their color or pattern, depending on the time of day. In general, camouflage which depends only on color becomes ineffective in the dark. Chameleon prawns, for example, which are opaque and patterned so as to blend in with the weeds in daylight, become transparent at night.

Then there are all the organisms you don't see or don't much notice during the day, but whose feeding habits lead them out into the open at night. For instance, the feather stars, those delicate varicolored multi-armed echinoderms, move to such prominent vantage points as the edge of fan corals, the better to strain the currents for plankton. Also typically at night, crustaceans are at their most active. Crabs of all sorts emerge to entertain.

Lobsters, some of them brilliantly colored and patterned, may be trapped in the beam of your torch, their immense antennae waving about in uncertain irritation at the unwonted intrusion. During the day, the diver is lucky to catch more than a glimpse of many animals, perhaps no more than the twitch of antennae under an overhang. Some species of shrimp rarely seen in the day also appear at night, when their predators have gone to sleep up in secure rocky crevices. Mollusks are another group which are active at night. Snails are out and about, some of them scraping away, feeding on the algaes associated with the hard coral. On the coral wall, on the seafloor, all sorts of shellfish are at large, mobile to an extent most people never suspect.

Since nightfall, as well, the hard corals have been showing more color, tentacles extended to feed on passing zooplankton, shrimps, eggs, and larvae. The larger polyps even capture very small fish. On the night shift, then, it is the polyps themselves which bring home the bacon; their zooxanthellae cellmates cannot engage in food production till sunrise makes photosynthesis possible again.

Dawn: Songbirds in the forest back of the beach herald the first faint wash of color in the sky. Now the sun is floating up on the horizon far away over the sea to the east, its rays triggering a whining, buzzing cacophony of cicadas on the hill behind the lagoon. Rising higher, the sun touches the trees lower and lower on the hillside, progressively setting off larger and nearer choruses of these insects, a great hysterical wave of sound washing towards you as you prepare to make the first dive of the day.

Things are also getting noisier beneath the sea. The relative quiet of the night has given way to a steady background of crunchings and clicks and pops. Sharp snaps suggest there are pistol shrimp about. The parrotfish are also awake and, underwater, you can hear them feeding on the hard corals. Their extremely hard, platelike teeth scrape at the coral skeleton to get at algae and the resident polyps, most of which have already retreated into their limestone cells for the day. (Much of those white sand beaches we all so admire are in fact the remains of parrotfish meals beyond number; it has been estimated that a single good-sized parrotfish produces around 16 kg of sand per year.)

At dawn the night-resters are hungry; and the reef is suddenly alive with fish; nocturnal, diurnal, and crepuscular feeders are all out and about at the same time. (Most large inshore fish are crepuscular feeders, feeding at dusk and dawn. This is a good time to photograph these species, since they are less easily agitated at this time and thus easier to approach.) All the predators which hunt by sight are now awake and going about the day's business. One of these is the long and slender trumpetfish, which at night snoozes vertically among the whip coral from which it is then virtually indistinguishable. With daybreak, however, the trumpetfish is ready to feed. A master of deception, it now adopts a new device. Following a grazing parrotfish as it goes about nibbling at the coral, the trumpetfish aligns its body closely with that of the larger fish, following its every move exactly, using the innocuous coral feeder as cover while it stalks its prey. The trumpetfish's intended victim, unalarmed at the proximity of the parrotfish, at the same time allows the masked predator to get close enough it can launch its sudden sneak attack. Imagine you were having a picnic in a peaceful meadow, one sunny morning, and, unbeknownst to you, you were being stalked by some peculiarly long-legged leopard using a cow as cover.

The squirrelfish and soldierfish, their red coloration no longer affording protection, retire to caves and crevices where the diver can peer in to find them idling about, big, electric-blue-ringed eyes apprehensive of visitors, waiting till night falls again.

Dayshift at the cleaning stations finds new attendants and new customers both. The cleaner shrimps are tucked away out of sight; and small fish, cleaner wrasse, instead perform strangely agitated little dances to advertise their services. At the moment, a fat grouper has in response signaled its willingness to accept the attentions of the wrasse, while a grunt hangs back in queue, waiting its turn. (The moray would rarely be seen getting cleaned in daylight; it is snugged up securely in its daytime hidey-hole on the reef.)

Daybreak also changes the behavior of smaller fish. They begin to school in the morning. Schooling reduces the effective threat from fish who hunt by sight, since these predators usually go for individuals, and seem confused by the choice a school offers.

This confusion does not apply to sharks, however, which in the spawning season may be seen feeding on swarms of squids, ripping through them like voracious scoop shovels, squids hanging from their jaws on all sides. (Of course sharks hunt partly by scent.) Nor are lionfish subject to confusion at the sight of a school; they merely hang there mimicking clumps of floating weed, waiting till some candidate for lunch swims too close, at which point their lower jaws drop open and the victims are sucked in by the sudden vacuum.

Midday: The morning advances, and the larger fish go off to deeper waters until the dusk feeding on the reef, permitting the tiddlers to relax a bit. As the sun rises higher and higher, the visibility improves. This is not simply a function of there being more light—most of the zooplankton head deeper to spend the daylight hours just off the bottom or in the sand itself, clearing the water. It is now that you get the best colors in shallower water, the most astonishing vistas.

Following the plankton, the feather stars are now generally found under plate corals and rocky overhangs, continuing their hunt for nourishment. Many of the hard corals, meanwhile, have withdrawn their tentacles, the night's hunting of plankton at an end. Primary food production has been turned over to the zooxanthellae, with conditions now ideal for photosynthesis.

Dusk: Late in the day, as the light fails, the plankton start to rise again. Once more, as there was at dawn, there is an intermingling of species normally found segregated in the daytime. Many holes and crevices on the reef witness an exchange of residents, much as if the same apartments were being rented to two different groups of tenants, the nightworkers and the dayworkers. Larger fish return to the reef to take advantage of unwary commuters during the sleepy changeover from dayshift to nightshift. As it becomes too dark for sight-hunters to operate, the great streaming banks and clouds of smaller fish gradually disintegrate, with single specimens and small formations of individuals now being the rule, many of the species in evidence predominantly red in coloration. The sea anemones have closed up, great orange and purple animal flowers with their resident clownfish swaddled securely within their stinging tentacles till morning. As the crustaceans and mollusks venture out to scavenge and the parrotfish go to sleep, silence descends with the night.

You turn off your torch and look up to where a crescent moon is shattered on the gentle ripple 12 meters above you; then you ascend the anchor line to the boat.

*In the daytime, Moorish idols (*Zanclus cornutus*) are strikingly patterned, and are often found engaged in stately spiralling dances involving as many as a dozen individuals. At night, the vividly contrasting bands turn more subdued, while Zanclus behavior changes. For one thing, after dark they tend to be solitary. (***Above:** *Koh Similan, Similans, in day, in 18 meters.* **Below:** *Koh Similan, at night, in 24 meters.)*

Foliose or 'rose' corals such as this one (Montipora sp.) *belong to a scleractinian genus which is part of the family Acroporidae, which includes the genus Acropora. Montipora are generally found in dense groups; this stand was growing on a reef slope off Koh Payu in about 24 meters. These corals are favorite targets for crown-of-thorn (Acanthaster planci) attacks.*

This scene was recorded off Koh Similan in 30 meters. A dense school of baitfish forms a halo over thriving plate corals and soft corals.

Another scleractinian, Echinopora lamellosa *is sometimes called hedgehog coral. It takes a lettuce-like, foliose form. (Koh Similan, Similans, in 15 meters.)*

Pavona decussata *is another foliose species. These corals were growing in 12 meters off Koh Ngam Yai, Chumphon.*

Chapter III

The Characteristic Shape of a Fringing Reef

Before looking at how the reef itself gets its characteristic shape, we may begin by asking how any given individual coral colony takes the form it does.

The shape of an individual coral colony. Brains, tables, mounds, domes, shrubs, vases—these shapes reflect a variety of adaptations to environmental factors such as the force of wave and current, the availability of light, and competition from other species.

First of all, of course, the animal colony follows genetically prescribed patterns of growth. Just as a puppy will not grow into a cat, a typically branching coral will not develop into a boulder-like shape. A genus such as *Acropora* will tend to construct the branching structures typical of staghorn, elkhorn, and shrub corals, while *Porites* will usually form boulder-like shapes. The various species have evolved the range of shapes they display in order to maximize exposure to light and food-bearing currents, while resisting the force of waves and current. They have adopted the features they have to adapt to one or the other of those conditions in different proportions, depending on the nature of their different preferred niches on the reef.

Beyond that, however, and unlike the case of the puppy who is very unlikely to develop cat-like features, local conditions can modify these patterns: mounds can become sheets or crusts, convoluted shapes may be transformed into tabular, while branching forms can become knobbly or plate-like. For example, energetic wave action or strong currents on an exposed seaward edge would quickly break up the delicately branching coral often found in relatively undisturbed water. The same species, however, where it is constantly buffeted by waves, grows in a markedly denser, more compact manner. Or, in deeper waters, where there is less light, we find the staghorn adopts a different shape again—this time coming to resemble plates or tables.

A species such as *Turbinaria mesenterina* also adopts a different shape, depending upon whether the colony develops in relatively shallow or in deep water. In shallow, muddy inshore environments, for example, it adopts a convoluted shape which can better take advantage of the abundant sunlight, with its zooxanthellae arrayed in all positions so that a good number of them may catch the direct sun no matter the time of day. In deeper water, though, it looks like an entirely different species: now its spreading plate-like form displays more polyps per unit of surface area, seeking maximum exposure to the weaker sunlight. (Marine scientists have transplanted convoluted varieties from shallower to deeper waters, and found that the corals changed their pattern of growth gradually to assume the plate-like form.)

Reproduction among the hard corals has two dimensions. On one hand, the individual colony grows by a relatively simple process of asexual budding (see Figure 4). On the other hand, dispersal—the establishment of new colonies—employs other means, both sexual and asexual. Corals have an interesting sexlife.

Many species, for example, are 'sequentially hermaphroditic', a practice which is no doubt illegal in some places. Some colonies, then, are one sex up to a certain age and afterwards they change. As though even such variety were not enough, the parent colony reproduces in two ways: through asexual budding, which, as we have seen, results in growth of the parent colony itself; and through the fertilization of eggs and subsequent dispersal by means of free-swimming larvae. In most species, the eggs are extruded and then fertilized externally. In a few others,

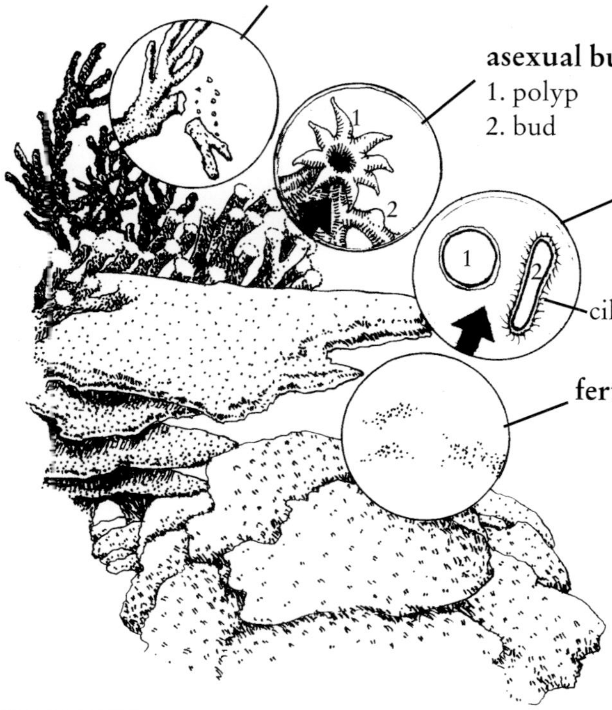

Figure 4: Reproductive strategies

the eggs are fertilized internally; the larvae develop in the 'stomach', and, when sufficiently mature, they exit through the 'mouth' of the parent. (One hesitates to refer to this interesting individual as a 'mother'. The mouth also serves as an anus, just to make things more interesting still.)

The larvae have cilia, little whip-like oars which are used in swimming, thus allowing for dispersal of the species to new areas. In this so-called 'planula' stage, the larvae are part of the plankton, and fair game themselves for hungry coral polyps and other predators everywhere. This phase may last as little as two days. Then they swim to the bottom and attach themselves, preparatory to metamorphosing and taking up life as the founders of new colonies of hard coral.

On the Great Barrier Reef, for example, the moon periodically inspires a mass coral spawning. Over a six-month period, sex organs mature and eggs ripen. These processes accelerate with the coming of warmer waters in the spring and the appearance of a full moon followed two days later by the spawning, which peaks on the fourth, fifth, or sixth night. The corals produce great quantities of eggs and sperm—the waters are so full of it even the most gluttonous of predators are defeated, finally, and sufficient fertilized eggs survive that the preservation of the species is ensured.

Mass spawning on this scale has yet to be observed in Thai waters, though during the dry season (January-February) many coral species do spawn at the same time.

Why do so many corals, including closely related but different species, spawn all at the same time? Normally, evolution would avoid hybridization, so these various species would reproduce at different times. Another side of the mystery: how do the sperm of a particular species find their appropriate mates? There must be compelling reason, from an adaptive point of view, for this peculiar approach to reproduction, but so far it remains an enigma.

Dispersal also takes place asexually, at least in some species. Storm damage, for example, may break tips off branching corals and tumble them down the reef slope or even carry them some longer distance away to generate new colonies, thereby turning apparent adversity to advantage. And asexual reproduction has advantages over sexual, in some respects: firstly, dispersal can take place any time of the year; secondly, the fragment can skip the vicissitudes of the

This coral, found in 18 meters off Koh Bon, Similans, is a boulderlike variation of Faviidae. *Massive colonies can be hundreds of years old, growing to five meters in height.*

Psammocora digitata, *or 'exclamation coral', is usually found in its massive, columnar form. (Hin Pae, Chumphon, in 12 meters.)*

Found in 14 meters off Koh Ngam Yai, Chumphon, this massive 'brain coral' (Lobophyllia sp.) *has lobed cups over its surface.*

Mushroom corals (Fungia sp.) *are often found sitting loose on a sand bottom, generally in three meters or more of water. (Koh Ngam Yai, Chumphon, in 7 meters.)*

Above: *Acropora corals often appear in dense stands over large areas of reef. More vulnerable to storm damage than massive species such as* Lobophyllia *or* Porites, *the branching corals compensate by typically growing at a faster rate, taking advantage of the abundant light and turbulent water, rich in gases, found in relatively shallow water on the seaward reef edge and slope. In terms of branch-tip extension, some species grow as much as 10 cm/year. (Koh Similan, Similans, 9 meters.)* **Below:** *Referred to variously as 'table corals', 'staghorn' or 'branching corals',* Acropora sp. *colonies can assume a variety of shapes, though in general they are circular with a branching network. The deeper they are found, the more they adopt a tabular or platelike shape, thereby maximizing exposure of polyps and associated photosynthetic algae to sunlight. (Koh Bon, Similans, in 12 meters.)*

planktonic stage; and, thirdly, the new growth is relatively mature already and can survive more surely in less specialized environments.

The shape of the reef. As we have seen, the living coral is merely a thin skin over the skeletal matrix already deposited by the colony. Similarly, the living reef is no more than a thin veneer over older reef formations.

The gross differences in shape between one reef and another may often be attributed to erosion—to forces which carved up the underlaying reefs when sea levels were lower and these formations were exposed to the waves. Despite these large-scale differences, however, the structure of reef communities everywhere is in many ways strikingly similar.

The three main types of formation are usually described as fringing reefs, barrier reefs, and atolls (see Figure 5). Fringing reefs, which are all that are found in Thai waters, develop close to shore, and have flat platforms extending out from the shore to raised rims and then abrupt drop-offs into deeper water (see Figure 6). Barrier reefs display much the same features, but are found farther from shore (160 km from shore, in some places along Australia's Great Barrier Reef). Thus there is a lagoon or deep channel separating the reef flat from the shore. It appears that a fringing reef is the basic type, and that it becomes a barrier reef, for example, when the land mass subsides, opening up a wider channel between reef and shore. An atoll, on the other hand, is a circular or horseshoe-shaped formation around a central lagoon. This version is simply a fringing reef whose island has sunk. This account of things, first proposed by Charles Darwin, may be modified by Daly's 'glacial control theory', which suggests that rising sea levels associated with the retreat of the icecaps has often led tropical reefs to grow apace, with the living corals always keeping close enough to surface for photosynthesis to take place.

In any case, fringing reefs, barrier reefs, and atolls are all of a basic type, in terms of biological form and the processes by which they grow.

First of all, similar limits on growth are enforced wherever the reef develops. The upper limit on vertical growth is imposed by the low tide mark, since the coral will die if exposed too long to the air. The lower limit lies about 50 meters beneath this level, since stony corals rely on sunlight, and at depths greater than this very little reef-building activity is possible (optimum conditions, in fact, extend from just beneath the surface to about 30 meters).

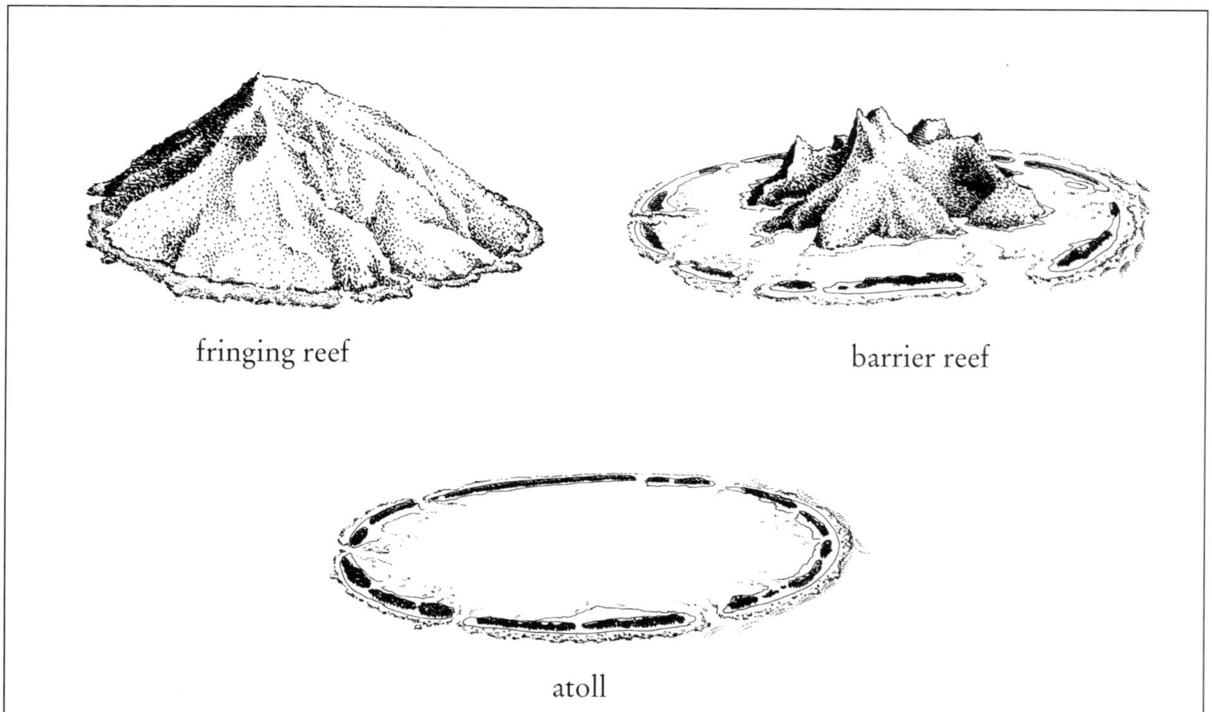

Figure 5: Three main types of reef formation

A reef is made up of many coral species and many individual coral specimens. The reef grows as the various colonies enlarge themselves and new ones are established. The individual colonies are cemented together by coralline algae and chemical precipitation of carbonates; grains of sand which are produced and distributed by a variety of natural forces and predatory organisms provide fill for the mortar. Sponges, for example, have important roles in reef-building. One of these roles, ironically, lies in the fact that some species bore into the coral, in the process destroying part of the hard matrix, and weakening branching corals in particular to the point they become vulnerable to storms. The broken coral, coarse and fine, settles in holes and crevices and is eventually cemented to help constitute the rocky reef substrate, affording greater protection from really destructive forces such as typhoons. Thus, boring sponges destroy only about 10 percent of the coral limestone through chemical dissolution, while the rest is redistributed and utilized with the neat economy characteristic of the rest of this ecosystem.

Growth rates vary: some staghorn corals (e.g. *Acropora cervicornis*) grow by as much as 10 cm/yr, in terms of branch-tip extension. Dome corals (e.g. *Montastrea annularis*), on the other hand, increase in height by only 0.25 to 0.70 cm/yr. Overall, in any case, the annual accretion of the reef probably ranges from 5 mm to 20 cm. (One square meter of coral polyp surface can produce about 10 gm/day of limestone.) The processes of calcium carbonate deposition by corals, algaes, and mollusks, together with cementation of hard debris, it is believed, have kept pace with rises in sea level due to glacial retreat. Adding to the substratums laid down in earlier epoches of global warming with their associated rises in sea level, this process has in some parts of the world resulted in reefs hundreds of meters thick. (An interesting thought: if the coral reefs were to die off, many Pacific coral islands would disappear under rising sea levels expected by many in the coming decades. At the rates most often predicted for this rise, however, living coral reefs would keep pace, enabling the islands to keep their heads above water, as it were.)

As we shall see in the next chapter, coral reefs characteristically develop more or less distinct zones, the atolls and barrier reefs more so than the fringing reefs.

A coral trout (Cephalopholis miniatus) *against a collection of antipatharian, soft, and hard corals. (Koh Payu, 15 meters.)*

A shoal of baitfish swarm around some brilliant soft corals.

Chapter IV

A Swim from Beach to Reef Bottom

The day is heating up, and the cicadas are a buzzing urgency in the trees behind the beach. It's a pleasure to enter the cool and relative quiet of the water.

Different organisms adapt to different positions along the gradient of light, wave action, and exposure to sediment which extends from the shore to the bottom of the reef. Characteristically, then, a more or less distinct pattern of zones arises.

Closest to the shore—to the beach or to low limestone cliffs—there is the reef flat, where the accumulation of sediments, the reduced circulation of water and concomitant deficiency in dissolved gases, together with the periodic emergences at low tides limit coral growth (especially where exceptionally low tides coincide with strong seaward winds). This area tends to be self-sufficient, consuming what it produces and no longer contributing to reef growth.

Wearing a mask and snorkel, and swimming out from the beach at low tide, your face is only tens of centimeters away from the bottom. The refraction of the water increases the apparent size of everything by 25 percent, furthermore, so you seem in danger of scraping your nose, so close does the bottom appear.

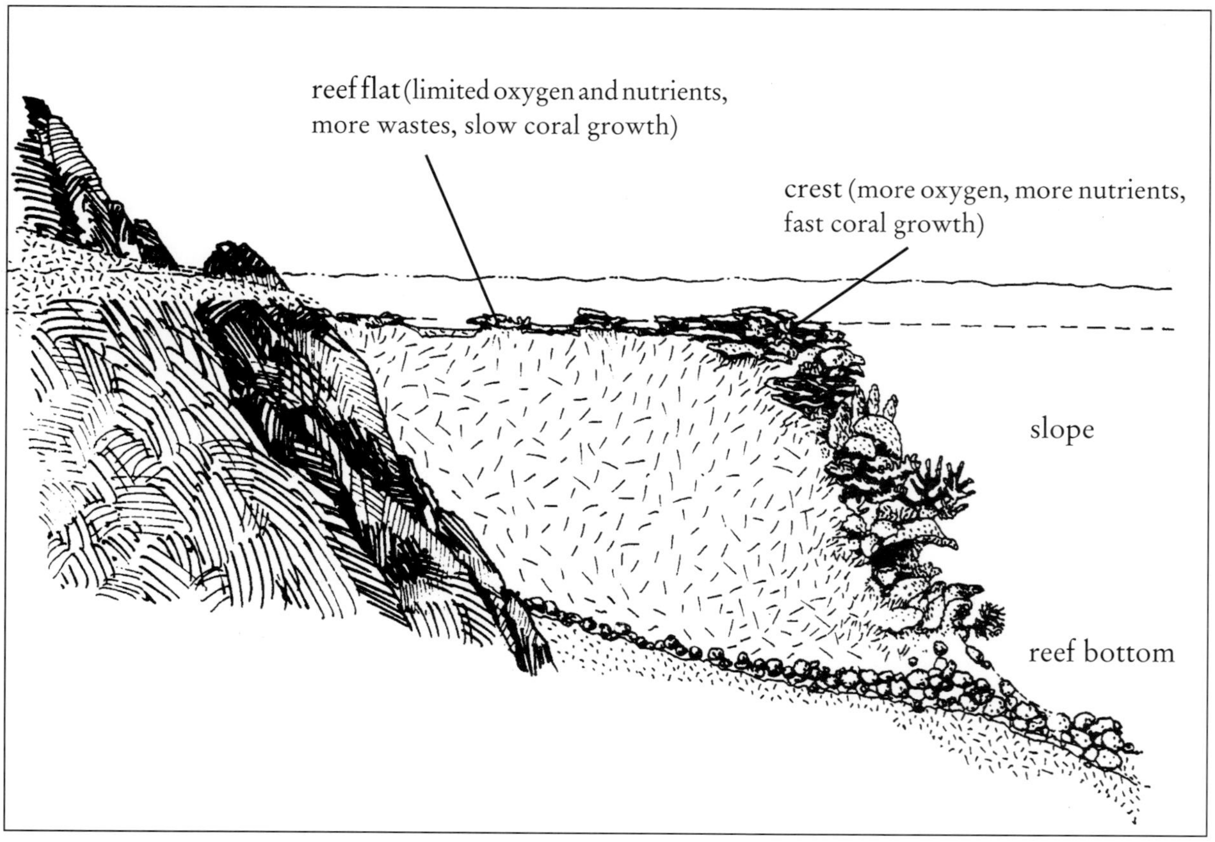

Figure 6: The shape of a typical fringing reef

Some uninformed snorkelers never get farther than this, and come away with a seriously impoverished notion of what there is to be seen in these waters. Clumps of weed and bits of dead and broken coral litter a muddy bottom; big black sea cucumbers loll about like gigantic worms; an occasional tiny reef fish flashes promise of what is to come, if only you explore farther.

Abruptly, the water deepens somewhat, and the relative wasteland of the flat gives way to compact, flattened colonies of coral, patches of live color, small branching corals, underwater 'pools' of vivid fish drifting and darting. These fish are bigger, more numerous, more varied. Anticipation grows as you proceed a couple of meters farther to where underwater ravines plunge off the seaward slope into deeper water still. Now you dive into a zone amazingly rich in fish, vast streams and schools of fish, brilliantly hued, larger specimens cruising through alone or in formations of several at a time. The variety and size and colors of the different corals is staggering.

Coral growth is mainly on the seaward edge and slope, so expansion of the reef is seaward, with the moribund flat gradually extending ever farther from shore. The upper seaward slopes are usually the most productive parts of the system, with most of the basic food production and deposition of the limestone skeleton taking place here. The series of ridges and ravines—or, as marine scientists term it, the 'spur-and-groove' topography—features buttresses of coral growth which project seawards, alternating with channels. These break up the force of the waves along the reef front, at the same time they provide lots of surface area for coral growth. (It is still uncertain how much of the spur-and-groove construction is due to shaped coral growth and how much is erosion from wave-tumbled boulders and sand. Most likely it results from a combination of the two.)

Here you find vigorous colonies of staghorn and shrub corals taking advantage of the abundant light and the turbulent water, rich in dissolved gases. These growths furnish both temporary shelter and permanent homes for a great variety of species. Swarms of little pastel coral fish drift in and out among the branches; tiny crabs retreat out of sight in the stony thicket as you approach. Fishes dart in to feed on filamentous algae, coral polyps and, of course, each other.

Farther down the slope, in deeper waters, plate corals as well as boulders and domes of coral appear more often. These more massive, more enduring forms can be hundreds of years old and as much as five meters high. Just as the branching corals host a variety of other organisms, these massive species also offer protection to a surprising assortment of worms, snails and shellfish. Branching corals, various species of *Acropora*, for example, may still be found at these deeper levels, but they tend to assume plate-like or tabular shapes now, maximizing exposure of their zooxanthellae to the weaker rays of the sun.

Everywhere on the seaward slope coralline algae grows, often resembling nothing so much as pink or red stains on the underlying coral or rock, yet this unassuming organism is an important reef builder, cementing hard debris and thus contributing to the integrity of the base structure.

Down towards the base of the reef, caves and overhangs provide refuge for soldierfish, squirrelfish, boxfish, and lots more. Many of these species are red in color, an admirable form of night camouflage underwater, and an indication that these fish are night hunters waiting for sundown before they get out and about. The overhangs in some cases are the result of coral growing outward in search of light. Caves and overhangs both, however, can in some parts of the world also be the result of wave erosion during an earlier epoch, when sea levels were much lower.

Lower still, at depths where many stony corals can no longer thrive, you encounter sea fans, sea whips, and black corals. Eventually, if you dive deep enough, beyond the range where there is enough light to support photosynthesis, you cease to find living hard corals.

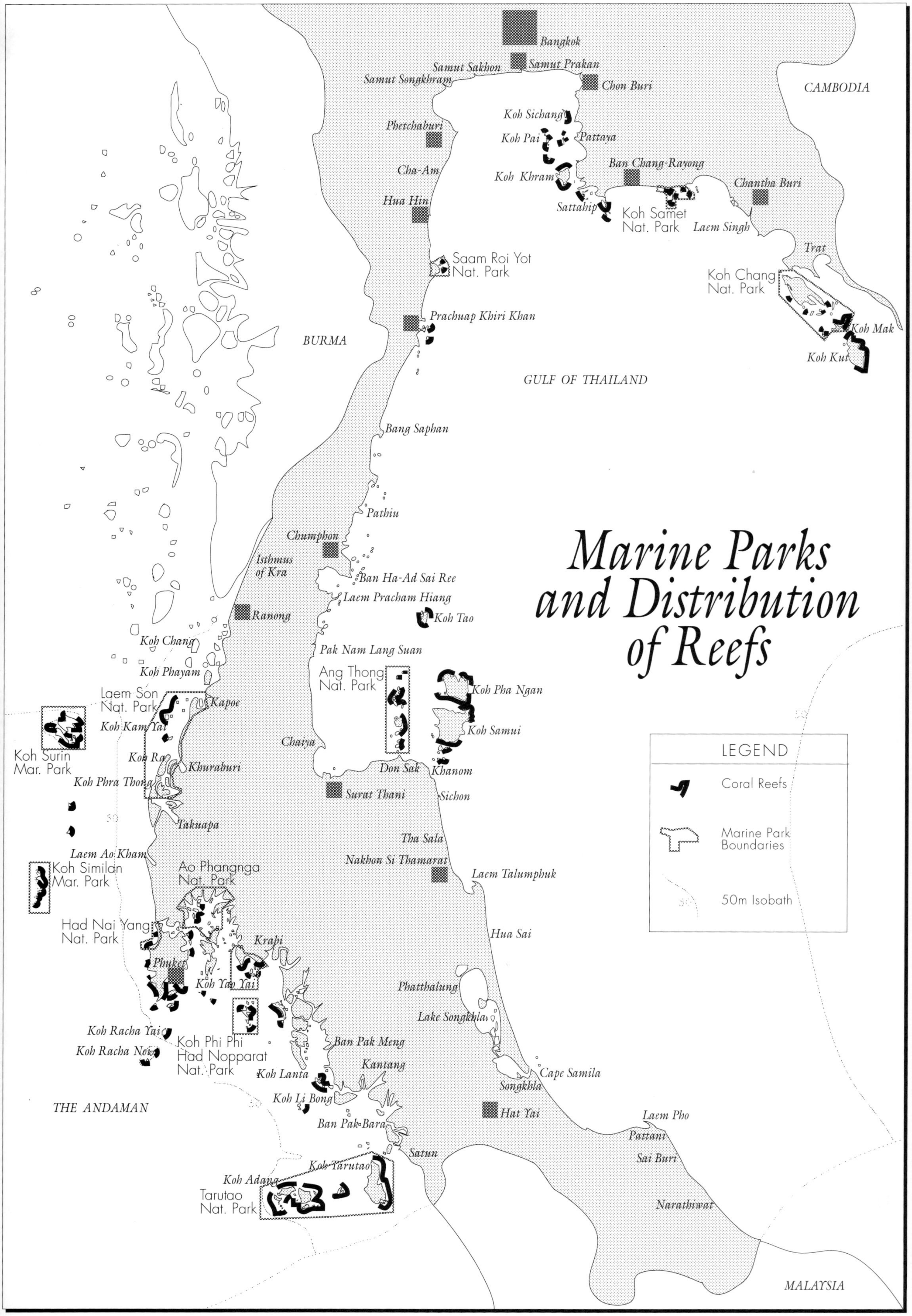
Marine Parks and Distribution of Reefs
LEGEND
Coral Reefs
Marine Park Boundaries
50m Isobath
Bangkok
Samut Prakan
Samut Sakhon
Samut Songkhram
Chon Buri
CAMBODIA
Koh Sichang
Koh Pai
Pattaya
Phetchaburi
Cha-Am
Koh Khram
Ban Chang-Rayong
Chantha Buri
Hua Hin
Sattahip
Koh Samet Nat. Park
Laem Singh
Trat
Saam Roi Yot Nat. Park
Koh Chang Nat. Park
Prachuap Khiri Khan
BURMA
Koh Mak
Koh Kut
GULF OF THAILAND
Bang Saphan
Pathiu
Chumphon
Isthmus of Kra
Ban Ha-Ad Sai Ree
Laem Pracham Hiang
Ranong
Koh Tao
Koh Chang
Pak Nam Lang Suan
Koh Phayam
Ang Thong Nat. Park
Laem Son Nat. Park
Kapoe
Koh Pha Ngan
Koh Kam Yai
Koh Surin Mar. Park
Chaiya
Koh Samui
Koh Ra
Khuraburi
Don Sak
Khanom
Koh Phra Thong
Surat Thani
Sichon
Takuapa
Tha Sala
Laem Ao Kham
Nakhon Si Thamarat
Koh Similan Mar. Park
Ao Phangnga Nat. Park
Laem Talumphuk
Had Nai Yang Nat. Park
Hua Sai
Krabi
Phuket
Koh Yao Yai
Phatthalung
Lake Songkhla
Koh Racha Yai
Koh Phi Phi Had Nopparat Nat. Park
Koh Racha Noi
Ban Pak Meng
Kantang
Koh Lanta
Cape Samila
Songkhla
Koh Li Bong
THE ANDAMAN
Hat Yai
Laem Pho
Ban Pak Bara
Pattani
Satun
Sai Buri
Koh Tarutao
Koh Adang
Tarutao Nat. Park
Narathiwat
MALAYSIA
50

PART TWO

THREATS TO THE CORAL REEF

*Dawn's early light sparkles on surface as the reef comes awake. This gorgonian coral (*Rumphella sp.*) has a thick corky sheath containing photosynthetic algae, and it thrives at depths and in areas with lots of light.* Rumphella *are among the strongest varieties of gorgonian, their external sheaths and wiry skeletons of gorgonin allowing them to grow in zones close to the surface where not only is there plenty of light, but where water movement is too violent for most other gorgonian species. (Koh Hu Yong, Similans, in 12 meters.)*

Chapter V

Natural Pressures

Of course all is not benign coexistence on the reef. These are not static communities: aside from continuous erosion, natural catastrophes of one sort or another must be accommodated. On one hand, grazing, predation and competition between coral and other species is constantly shaping the community as a whole. Meanwhile, space for new growth is from time to time opened up by destructive forces in the environment. Some of these forces are simply the result of changes in the limiting factors which permit coral growth in the first place.

Natural changes in limiting factors

Conditions of light, temperature, salinity, and wave energy determine where and when a reef can develop.

Light: A relatively pure oceanic reef, well offshore and in very clear water, relies almost entirely on photosynthesis; and it is sunlight which provides the basic energy to build the reef. Light is essential, then, and since light is a function of both depth and turbidity, limiting factors are immediately evident. In clear water, 50 meters is the approximate cutoff point for most hard coral growth (optimum conditions extend only to about 30 meters); after that, there isn't enough light for the zooxanthellae to produce required nutrients. As one indication of how dependent the hermatypic scleractinian corals are on light, their normal calcification rates may be halved on a cloudy day.

Any event which raises the water level, consequently, will tend to kill off the deepest corals. At the same time, anything which increases the turbidity of the water reduces the depths at which the reef can grow, and inhibits hard coral growth at all levels. In the natural course of events, the advance and retreat of glaciation has been the main cause of fluctuating sea levels, with the seas having been at times both somewhat higher than they are now and more than 150 meters lower. Another natural phenomenon which has reduced available light to the point reefs have suffered is vulcanism, where volcanic eruptions have been so violent and prolonged that the sun has been blacked out. (Humankind now has the opportunity to engineer our own version of this event, given the possibility of a 'nuclear winter'.)

Temperature: Temperature is the next crucial factor. The hard corals can grow only within a certain range of temperatures (higher than 23°C for well-developed coral reefs; though there are reefs in the Florida Keys which have developed at temperatures as low as 18°C, and some scleractinian corals can grow in communities, as opposed to reefs, at temperatures as low as 14.5°C). For this reason, coral reefs are restricted to regions which fall approximately between latitudes 23.5 degrees north to 23.5 degrees south, within the limits of the sun's winter and summer elevations. Thai waters are perfect, with annual water temperatures never falling below 28°C on the Andaman Sea and 26°C in the Gulf.

As of this writing, sea temperatures have recently been rising worldwide, and significant 'bleaching' of corals (where, under stress, the zooxanthellae are leaving the polyp hosts) has been reported from reefs around the globe, including those in Thai waters. Whether this is a result of a supposed global greenhouse effect, or whether it is due to some shorter-term phenomenon, remains to be seen.

Salinity: The stony corals generally require open-ocean levels of salinity, approximately 30-38 parts of salt per thousand parts of water (the salinity on Persian Gulf reefs can reach 40 ppt). Coral reefs have a hard time establishing themselves in regions close to river mouths, for that

*Sea temperatures have been rising worldwide, and significant 'bleaching' of corals has been reported from reefs around the globe, including those in Thai waters. Apparently the stress of higher temperatures is causing the symbiotic zooxanthellae to leave their polyp hosts, leaving many corals without adequate nourishment. Although these shots were taken in the Gulf of Thailand, bleaching was also apparent in the Andaman Sea, affecting both hard and soft corals as wells as anemones and sponges. Whether this bleaching resulted from a global greenhouse effect, or whether it was due to some shorter-term phenomenon, remains uncertain. (***Top, left:** *Koh Ngam Yai, Chumphon, in 9 meters.* **Top, right:** *Hin Pae, Chumphon, in 9 meters.* **Below:** *Koh Ngam Noi, Chumphon, in 9 meters.)*

Competition from other species is one constant pressure on most reef organisms. Here, either a staghorn coral is growing in a barrel sponge (Petrosia sp.) or the sponge is growing around the staghorn. Meanwhile, black damselfish (Dacyllus trimaculatus) *hover over the whole. (Koh Miang, Similans, in 22 meters.)*

In 1989, this patch of staghorn coral (Acropora robusta) *in 10 meters of water off Koh Ngam Noi, Chumphon, was in fine shape (**before, above**). Then Typhoon Gay struck.*

Though this freak typhoon was more violent than most local tropical storms, storm damage is one common natural pressure on coral reefs. Two years after Typhoon Gay had passed, the staghorn was still completely dead.

Two or three weeks after a crown-of-thorns infestation has bleached out an area of reef, fuzz-like filamentous algae starts to grow on the bare patches. A few months later, typically, encrusting calcareous algae and soft corals begin to make their appearance. After several years, the area sees new colonization by hard corals, slow at first to gain a foothold on the loose coral rubble carpeted with algae. First the faster-growing coral species, then the slower, establish themselves until, after 20 to 40 years—supposing other pressures have not interfered—the reef is back to where it was before the crown-of-thorns laid it waste. (Hin Pae, Chumphon, in 12 meters.)

A common sound underwater, around coral reefs, is that of parrotfish such as this Scarus frenatus *grazing on hard corals, using their beaklike teeth to scrape at and crush the exoskeleton to get at the polyps inside. (Koh Similan, Similans, in 9 meters.)*

One more natural pressure on the corals is that of competition between coral species or coral colonies themselves. They can inhibit one another's growth in two ways: where one colony doesn't simply overshadow another in the competition for sunlight, it may instead extend its mesenterial filaments through the mouths of its constituent polyps and digest its neighbor. Here we see Galaxea sp., *a hard coral, attacking another* Galaxea. *(Hin Pae, Chumphon, in 12 meters.)*

Above: *Although clown triggerfish* (Balistoides conspicillum; *Koh Similan, in 18 meters) are not very common around the reef, they are conspicuous once spotted. Coral grazers, they defend themselves with a sharp dorsal spine which springs erect and locks, making them difficult to extricate from among the corals. This has proven little defence against spearfishers and commercial aquarium fish collectors.* **Left:** *Egg cowries* (Ovula ovum) *are getting rarer on Thai reefs. The elegance of their shells, comparable to finest porcelain and irresistible to collectors, threatens to doom them locally. (Koh Surin in 15 meters.)* **Below**: *Related to seahorses, ghost pipefish* (Solenostomus paradoxus) *frequent sea grass and algae beds, well camouflaged by color and skin appendages. Exotic reef fish are prized by commercial collectors, some of whom use poison to stun and catch these creatures. Many specimens die before delivery as a consequence. (Koh Similan, at night, in 12 meters.)*

*uman moves to redecorate the coral reefs are hardly an improvement on ature. (***Above:** *Koh Larn, Pattaya, in 11 meters.* **Above, *right***: *Koh Bon, imilans area, in 15 meters.)*

*single anchoring over a virgin reef can destroy a square meter of coral rowth, the anchor chain itself doing much of the damage, as the boat oves to and fro with wind and wave. (***Below:** *Koh Bon, north of the imilans, 12 meters.)*

reason. In other areas, freshwater runoff following a typhoon, for example, may reduce salinity enough that corals are affected. One reason the corals of the Red Sea, in the Middle East, are so spectacular is that the surrounding area is extremely arid, and the reefs are spared both freshwater runoff and the concomitant deposition of sediments.

Wave energy: Wave and sand erosion takes its constant toll on the reef, especially during storms.

Typhoons, though infrequent in Thai waters, can be a major cause of damage, sometimes tearing up large areas of coral. It may take affected reefs several years to recover from the destruction wrought by a storm such as Typhoon Gay, which struck the Chumphon area in 1989. Many branching and unattached corals were broken up, while some reefs were covered with sand and rubble.

Another threat, though an uncommon one, can be where strong seaward winds combine with exceptionally low tides to expose reef flats to the air for so long that coral dies of dehydration despite the protective coating of mucus that normally protects it.

Bioerosion

Many species of organism feed either directly on the coral, or else destroy the coral skeleton in the process of feeding on algae or various creatures which live inside the skeleton. (Coral species or colonies themselves frequently compete with each other, and they can inhibit one another's growth in two ways: where one colony doesn't simply overshadow another in the competition for sunlight, it may instead extend its mesenterial filaments—through the mouths of its constituent polyps, of course—and digest its neighbor.)

Parrotfishes, for example, use their hard plate-like teeth to scrape away at algae around the periphery of the live coral colony, gradually undercutting the coral head and in the process manufacturing prodigious quantities of gorgeous white sand per adult fish, thereby contributing much to the appeal of Thailand's beaches. Sea urchins graze the coral even more extensively. Where organisms such as the parrotfish, surgeonfish, urchins, and certain *Pagurid* crabs have scored the surface, algae, bacteria, and other organisms may then invade the colony and cause further damage. Fish of the genus *Arothron* nibble at the tips of branching corals such as *Pocillopora*. Snails such as *Jenneria pustulata*, taking the night shift, also prey on these corals. *Balistes polylepsis* bites off the protrusions on corals of the genera *Porites* and *Pavona* looking for the tasty snails, worms, and shellfish which live within the stony coral matrix. Certain aeolid nudibranchs not only derive nourishment from the corals, they also ingest the nematocysts, the stinging cells, and redeploy them as a defence system. Among the best-known and most destructive coral feeders are the starfishes. The crown-of-thorns (*Acanthaster planci*), for example, everts its stomach through its mouth and secretes the juices necessary to digest the coral polyps, leaving bleached white patches on the coral colonies wherever it has been.

Periodically, crown-of-thorns populations explode, causing widespread and severe damage. Many researchers, however, believe that this represents a natural pruning process, one ultimately beneficial to the reef, just as moderate grazing by the variety of other organisms has been found to be best for overall coral growth.

Studies show that the natural kinds of reef destruction referred to above are generally functional for the reef community in the longer run, dispersing the seeds of new colonies and reefs, making room for species diversity and new growth, and in some cases helping in the process of establishing and maintaining a sound reef substratum (where sand and other rubble fills in holes and crevices subsequently to be cemented by various reef-building organisms.) Under normal circumstances then, in natural history, the various forces of creation and destruction have been kept in a nice balance.

Chapter VI

Human Pressures

The more gradual processes of erosion—wave and sand action together with bioerosion—are generally in balance with reef growth, and are probably in fact necessary for the overall health of the reef.

A new element has entered the equation, however. Human beings are no doubt part of nature, but we are adding pressures on reef populations which are 'unnatural', and which are not sustainable.

Thailand is one country where there is still time to institute policies of economic development which would be conservative of the marine environment. This opportunity is fast disappearing, though, in the rush to exploit the nation's natural resources to the fullest possible extent in the short term, with little thought for what the future holds in store.

Pollutants

The human population is exploding at the same time an exploitative, wasteful, and careless technology spreads its promise of material progress to all parts of the globe. The consequences of this conjunction of factors are many, but one which is demanding greater and greater concern is the question of waste disposal. Mankind is soiling its own nest on a scale which would have been almost inconceivable only a few decades ago. The reefs are just a part of the larger environment which is at risk.

Tourism: Commercial tourism in Thailand is a leading source of damage to local reefs. At the same time the foreign tourist has been touted as an essential resource by government and the business community, insufficiently controlled development of the tourism industry is threatening the natural resources of the marine environment—resources with a much longer-term potential value than tourism. Before long—already in areas such as Pattaya—revenues from foreign visitors will drop off, since one of the main attractions of Thailand has been the beauty of its natural environment, and this treasure is being squandered.

Just one source of damage from tourism is pollution. And one of the most destructive forms this takes, especially around developed resort areas, is the disposal of untreated sewage into the sea. Untreated sewage can make the water not merely unpleasant but even dangerous for water sports. Pattaya Bay, long Thailand's 'premier beach resort' in the terms of its exploiters, by the end of the 1980s was no longer safe for swimming, given the levels of dangerous bacteria frequently found there. Neighboring Jomtien Beach and Larn Island are threatened with similar levels of pollution. The coral on Koh Larn has been all but wiped out through a combination of pollution and other factors.

Another kind of pollution, one which annually gets worse, is the dumping of rubbish by tour boats and their customers. This sort of practice, even if we overlook the aesthetic trauma of finding plastic bags and soft-drink cans everywhere on the reefs, is harmful to the corals and to some of the other reef animals. Shifting with the currents, it can abrade or break coral. There is also evidence that sea turtles, for example, already threatened by human predation and disruption of their nesting grounds, sometimes are killed by running afoul of rubbish. Compared to sewage disposal, this form of pollution may currently be less destructive, but its effects are seen over larger areas, with fringing reefs on some of the remotest islands already being affected.

Industry: Industrial wastes have by now polluted the entire globe. For years, autopsies

Above: *Trawlers snag their nets on coral reefs. Then, when the fishermer dive to free the nets, they do further damage to the coral. (Hin Pae Chumphon, in 15 meters.)* **Left:** *Aside from damaging delicate corals wit missed shots, too many spearfishermen decimate various reef fish species At the same time they condition other large fish to shy away from humans helping to spoil sport diving for those who come after. (Hin Pae, Chumphon in 9 meters.)* **Below:** *A fishing net covers a bed of sea anemones. (Ko Ngam Noi, Chumphon, in 12 meters.)*

This anchor is being dragged on a collision course with delicate branching corals, stands which took years to develop. (Koh Gra, Trat, in 7 meters.)

Coral rubble is all that is left of once-thriving colonies, at this spot off Koh Similan.

on penguins from Antarctica have revealed clearly discernible levels of DDT in their tissues, and oil slicks can be found on the remotest seas. As Thailand enters NIC-hood, becoming a 'Newly Industrialized Country' in the coming decade, this country will more and more come to experience the costs of industrial development. Ideally, its leaders will at the same time come to feel the responsibilities entailed—the responsibility to their own population and natural environment, as well as the broader responsibility to the world community.

In the meantime, untreated poisonous industrial wastes are being dumped into the seas and waterways. And there are other sources of destructive chemical pollutants. Freshwater runoff, as we have seen, can already have an adverse affect on corals by reducing the salinity of the water at the same time it dumps sediments on the living reef. To make matters even worse, however, they frequently carry insecticides (especially where these are being overused).

'Enriching' pollutants, on the other hand, include such things as sewage or fertilizers which, again, are washed down into the seas with the runoff. Normally the reef must carefully recycle and conserve scarce nitrates and phosphates. Nowadays, however, these nutrients may sometimes be too abundant—leading to algae and plankton blooms which increase the turbidity of the water to levels which seriously inhibit coral growth. Here we see ill-advised farming methods (over-use of chemical fertilizers and insecticides) having consequences far beyond those directly associated with threats to the health of farm workers, consumers, and the land itself.

Sediments pose another threat to Thailand's reefs. On one hand, increased turbidity reduces the sunlight available for photosynthesis. On the other hand, particulate matter in settles out of the water onto the reef and smothers the coral. The coral polyps need clear water in which to grow. In the area around Pattaya and Samae San, for example, there has been a marked increase in the turbidity of the seawater over the past few years due to the dredging operations for the new harbor at Laem Chabang and the land construction for the Eastern Seaboard Project.

Sports divers, for example, report a decrease in average visibility of from approximately 10-15 meters in the winter season 1985 to 7-10 meters in the winter season 1990; and year by year conditions continue to get worse. Residential and industrial waste disposal may increase the pressures considerably once the whole Eastern Seaboard Development Area goes into full production.

In the Phuket area, tin mining operations and the deep-sea port project at Laem Panwah have each resulted in some destruction of reefs. Mining operations, however, have in recent years been restricted to specific areas, and the damage is being contained. The damage caused by the port project, on the other hand, seems to have been only short-term. The dredging had resulted in a 30 percent reduction in coral cover in the affected area after one year; but recovery was apparent within two years of that. (It helped that the corals affected were hardy reef-flat species, ones naturally adapted to relatively turbid waters.)

Another source of sediments altogether has been exerting greater and greater pressure on the corals in recent years. The rape of one natural resource, the forests, is now threatening another, the reefs. Extensive deforestation is leading to extensive soil erosion, and that particulate matter is being washed down Thailand's streams and rivers into the sea. (Forest cover in Thailand has decreased from about 56 percent of the total land surface in 1960 to a mere 28 percent in 1990, where 28 percent is the figure used by the Forestry Department—figures as low as 13 percent for first-growth forest, estimates supposedly based upon satellite surveys, have been put forward by some foreign environmentalist groups.)

The latter problem is then exacerbated by extensive clearing of mangrove swamps, which are very good at trapping and holding such sediments, and this pressure on the reef habitat becomes a very serious matter indeed.

Studies have also shown that different coral species have different degrees of tolerance for sedimentation. The major builder of fringing reefs in Southeast Asia, *Porites lutea*, as well as another common intertidal species, *Montipora formosa*, are very resistant, being able to recover even after having been buried under sediments for some time. Staghorn coral (*Acropora formosa*), on the other hand, cannot be smothered by sediment because of its shape;

the increased turbidity associated with sediments, though, can very quickly kill it.

During the southwest monsoon season, wind and wave action may clean off some of the sediment which has settled on the reefs off Phuket during the northeast monsoon, allowing at least a temporary recovery of the corals. This indicates that if a program is instituted to improve conditions of turbidity and siltation, the reefs would be self-cleaning and the coral could be expected to regenerate.

Mechanical damage

Dynamite fishing: Dynamite fishing was widespread and common in Thai waters. Nitrate fertilizer mixed with diesel fuel and packed into condensed milk tins is sometimes used as the blasting agent; and there are rumors that really big bombs are used occasionally—huge earthenware jars normally used for storing water, packed with explosives. Charges are detonated at depths of 10-30 meters, often just off the reef slopes. This kills or stuns large numbers of the bigger fish which typically congregate around the reef at these depths, and which then float to the surface to be collected by the fishermen. (Though only an uncertain proportion of the kill is actually retrieved; the rest simply goes to the bottom and decomposes.)

This practice devastates the coral reef, in an instant leveling growth which has taken many years to establish itself. Large numbers of marine organisms of no immediate commercial value to the dynamite fishermen are killed. Even years later, in areas like these, there is often little more to see than a submarine desert—an expanse of bleached coral rubble, bits of staghorn, a few fish, sea stars, perhaps some soft corals. As a technique for fishing, dynamite blasting might be compared to a situation where hunters burned down entire forests merely to flush a few deer for food. However productive this method may at first be, from the fisherman's point of view, it tends to remove the very conditions which make choice fish gather in easily blasted congregations. Ever newer sites must be found for subsequent bombings, spreading the pattern of destruction farther and farther, while helping to ruin reefs for any kind of fishing.

Lack of manpower and other resources is in part what makes it difficult to enforce existing laws, since the reef-blasting occurs over large areas of Thailand's seas, many of them quite remote.

In any case, the practice is on the decline, and this in proportion to increasing tourism—the fishermen don't like to blast with people around. Unfortunately, the reefs still lose: the tourists are chasing the dynamite fishermen away, while at the same time they are bringing their own equally destructive pressures with them.

Other fishing methods: A less extensive form of damage associated with small-scale fisheries is the practice in areas such as Pha Ngan Island of tearing up live coral and using it to camouflage fish traps.

Or there is the technique referred to locally as *oo-un yippon*, or 'Japanese netting' (*muro-ami*, in Japanese). In the past, Sea Gypsies and others used to slap the surface of the water to herd fish, particularly open-water species such as rainbow runners, into nets. Now that there are fewer rainbow runners for the catching, however, fishermen actually pound on the reefs to chase all sorts of fish off and into the nets. In places, especially in some locations off the west coast of Phuket and off Koh Phi Phi, this is having a devastating effect on the corals.

Further damage still is caused by certain types of trawling, where the boats come in too close to the reefs, often despite existing legislation prohibiting this. Local fishermen have protested this practice by larger-scale commercial operations, sometimes violently, but with very limited success. (In April of 1990, following the shooting death of a crewmember on a trawler which was fishing out of season in the Gulf near Chumphon, the Fisheries Department finally asked the Interior Ministry to investigate the conflict between Pathiu District villagers and a certain "influential official in uniform" who allegedly was behind a large-scale illegal fishing operation.)

Anchor damage: Dynamite fishing is reportedly on the decline, but in many areas a new threat is contending for distinction as the leading cause of mechanical damage to the reefs. Tour boats, dive boats, and fishermen all tend to go to go to the same anchorages again and again. The traffic in tour boats taking visitors out to the reefs has exploded in the late 80s and early 90s. Naturally enough, the tour operators head straight for the best coral and then, quite thoughtlessly, drop anchor. At a rough estimate, a single anchoring over virgin reef destroys an average of one square meter of live coral. Dragging anchor chains are more devastating still. And, since broken coral takes several years to recover (a table coral one meter across can be ten years old) that damage is cumulative. On popular tourist sites, the ravages underwater have come to compare with the results of dynamite fishing.

Sports divers: If any group should already be educated in the need for conservationist attitudes, it is sports divers. Nevertheless, some divers are entirely irresponsible. They anchor on reefs; they trample and otherwise break up delicate coral growths with no thought of other people to come. Improperly weighted or merely careless, they may blunder about like bulls in underwater china shops. Often they aren't aware that even if you simply brace yourself against live corals, or hold onto them in the current, you may be disturbing their mucous coating and exposing them to bacterial and other damage. Just as visitors to forest parks are expected to leave things as they found them, explorers of our undersea gardens should show a similar respect for the environment and for the enjoyment of other people to follow.

Plunder

Commercial collecting: *Collecting* coral is illegal in Thailand. The possession or actual sale of coral, unfortunately, is not. Over 100 souvenir shops on Phuket alone openly sell shells and coral to tourists, who are generally unaware that collecting for sale has been officially banned.

Favorite hunting grounds for coral are the Racha and Phi Phi Islands, the west coast of Phuket, and the Similan Islands. *Acropora*, *Porites*, *Faviids*, and blue corals are the main types harvested. (Given how slowly corals grow, the development of sustainable commercial harvesting practices is unlikely.) Again, as with shells, local sale and possession is not illegal.

Shell collecting also results in considerable coral destruction, especially on the reef flats, with trampling of the live corals and deliberate overturning of coral heads during the search. The reef flats at Rawaii and Laem Panwah on Phuket, for example, have been extensively damaged in this way.

Traditionally, shellfish were taken by locals mainly for food, while trade in coral was limited. With the boom in tourism, thing have changed, and the reefs are fast being stripped. In the Andaman Sea, the Racha and Similan Islands have been the prime collecting sites. It is not certain how many locals are involved in the industry, but 400-500 Rawaii residents are at least part-time collectors, while undocumented but considerable numbers of people in other areas such as Phi Phi Don rely on collecting for part of their income.

The shells are sold to tourists and to wholesalers for export to Bangkok and abroad. The Coastal Resources Management Project estimated the export trade in Phuket shells was more than eight million baht in 1988, though of course the actual collectors themselves would have got much less than this. Already, several beautiful shell species, formerly abundant, have become rare. Varieties of cowrie (*Cypraeidae*) and cone shell (*Conidae*), as well as large giant clams and the green turban shell (*Turbo marmoratus*) have all but vanished.

Locals also fish reef species such as lobsters, groupers, sea bass, snappers, and grunts less for their own subsistence, now, and more for sale to restaurants and hotels catering to the tourist trade. Aside from its effects on the fish populations themselves, this factor has uncertain consequences for the reef ecosystem as a whole.

Aquarium fish are also collected, especially from the Racha and Similan Islands, with much of the catch going to foreign markets such as Hong Kong and Singapore. Poisons are sometimes used to stun and catch these colorful creatures, and many of the specimens die before delivery as a consequence. (It is estimated that there is an overall mortality rate of 80 percent from poisoning and other factors in the handling.) It is not yet certain which, if any, species are threatened with local extinction, nor has it been established what ecological implications such collecting may have. But given the collectors' frequent preference for juvenile fish, which survive better in the handling, it is likely the practice is having an adverse effect on breeding, at the least. In any case, there is already a discernible effect in terms of the loss to scientific investigators and sightseers alike. (The Phuket Marine Biological Center is trying to breed clownfish, one popular aquarium fish, in captivity, hoping that if the technique is successful it might lead to raising these and other fancy fish commercially, thereby taking the pressure off the wild reef varieties. Local environmentalists are talking about raising fast-growing corals for sale as well.)

Who can say what consequences uncontrolled collecting may be having on the marine ecosystem in the longer run? Some marine biologists have theorized that the large-scale commercial stripping of the spectacular giant triton shellfish from certain Australian reefs, for example, has resulted in highly destructive infestations of crown-of-thorns starfish which, with their main natural enemy gone, have been free to breed and prey on vast areas of live coral. (This hypothesis has been called into doubt, however, with the argument that the tritons were never really numerous enough to have been a significant controlling factor on crown-of-thorns populations.)

Souvenir collecting: The number of snorkelers and scuba divers which visit Thai waters is burgeoning. Of the more than two million visitors expected on Phuket in 1995, at least 7.4 percent (150,000) of these will be doing some snorkeling and diving. It doesn't take too many tour boats filled with divers to clean out most of the interesting shells and fragile corals, at least from the shallower waters. Whatever the relative ecological effects of this particular problem, untrained or irresponsible recreational divers are just one more pressure that could well spell the end of most of the reefs in Thailand.

Spear-fishing also takes a toll, and a ban on the activity should be enforced. There is very little sport in hunting reef fish with scuba gear and spear-guns—it really is shooting fish in a barrel. At least until most of the bigger fish are gone, that is; and until the larger fish which remain learn to shy away from human intruders (including divers who are just looking). Spear-fishing often results in coral damage, moreover, when hunters miss their shots.

Recovery times from natural pressures such as storm damage are quite good, where they have been measured in the Phuket area, for example. In places like Patong Beach, on the other hand, where there is damage from other causes such as anchors, trampling, and starfish predation, recovery has proven to be very slow. Local marine biologists suggest that it is the frequency of such damage which is the operative factor, here. There simply isn't time for recovery.

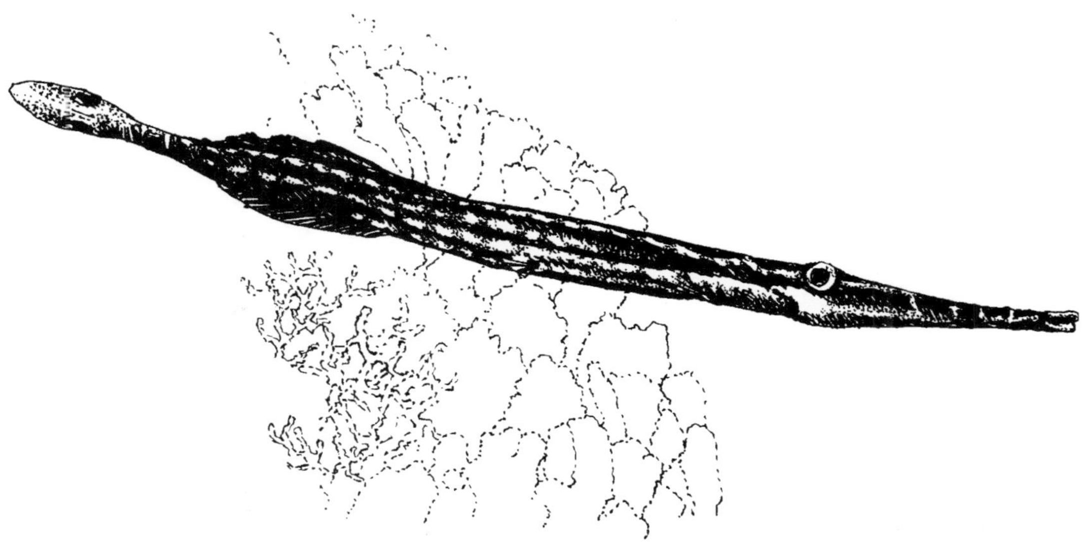

Chapter VII

Distribution and Condition of Coral Reefs

The Gulf of Thailand

The Gulf of Thailand in located in the inner portion of a geological feature known as the Sunda Platform. The Gulf resembles a big bowl with its center shifted slightly to the south. At its deepest it reaches 86 meters, though the average is about 30 meters. Temperatures range from 26-30°C; and salinity is 30 parts per thousand, except near the delta zone where freshwater runoff dilutes the sea. Throughout, however, the Gulf is less saline than the Andaman Sea. If you draw a line from Vietnam to Kota Baharu, in Malaysia, it describes an underwater ridge which tends to bottle up the Gulf, along with its load of freshwater runoff. Only the water relatively close to surface is readily exchanged at the boundary with the more saline water to the south. Above Koh Si Chang, there is no coral. Above Chumphon, there are only 'coral communities'. Further studies are needed to determine whether low salinity and turbidity related to freshwater runoff are largely to blame, or whether, as some opinion has it, the grazing of large numbers of sea urchins is responsible.

Many rivers run into the Gulf, and the sediment and fresh water they bring puts limits on where coral may establish itself. New and potentially much more extensive limits are being set, now, by tourism and coastal development projects.

In the innermost region of the Gulf, with four major rivers dumping sediments as well as (increasingly in recent times) chemical pollutants, there are no corals to be found. Coral reefs begin to appear at Hua Hin, in Phrachuap Khiri Khan Province, about 185 km from Bangkok. These reefs are mainly fringing the islands off the coast, where sediment deposition is less heavy.

More than 200 islands are found along the 1300-km west coast of the Gulf. These fall into three main groups: Phrachuap Khiri Khan Province, Chumphon Province, and Surat Thani Province which, with 80, boasts the largest group of islands.

Prachuap Khiri Khan: This area, 326 km south of Bangkok on the west coast of the Gulf of Thailand, includes many islands strung along 245 km of coastline. Both the diversity and the density of coral is low, especially in the more northern part, where only a few reefs are found; *Porites lutea* and *Acropora* predominate. Visibility is very poor, sometimes as low as 30 centimeters; the islands farthest from shore offer the clearest water.

The main problem here is sediment from freshwater runoff; what with the many rivers running into the sea carrying an ever-increasing burden of soil as deforestation strips the land to the north.

Farther south there are many more reefs, but here sediment being dumped by Klong Waan (Canal) has increased dramatically in recent years due to the clearing of mangrove swamps and the development in their place of coastal shrimp farms. A lesser source of damage has been reef blasting (reportedly this practice has decreased lately).

Chumphon: Farther south on the west coast of the Gulf, the Chumphon area has many coastal islands, with coral reefs on all of them. This has become a favorite destination for local sports divers in recent years because of the high species diversity and density of corals (various species of *Porites*, *Pavona*, *Acropora*, and more). Visibility is generally quite good here, still. In 1989, Typhoon Gay caused widespread havoc with the reefs, smashing delicate corals and burying others. Recovery is underway, however, at least if locals and tourists permit. Human damage has been mostly due to reef blasting and, more importantly these days, anchor damage and collecting associated with increasing tourism.

The ASEAN-Australia surveys have categorized the Chumphon reefs according to four types, depending on the nature and extent of damage suffered.

Samet Island (Chumphon) was an example of the first group, where heavy wind and wave action, together with turbidity from runoff sediments, has inhibited reef growth, so that only 'coral communities' are found, with isolated small massive corals and encrusting corals in evidence on the sand, rock, and algae bottom.

Khai and Kang Sua Islands are representative of the second group, which are far enough from the river runoff to be largely protected from siltation. Calm, clear waters permit a high degree of coral cover, with the fast-growing *Pavona descussata* predominating.

A third group of islands (e.g. Mud Wai Noi) have been well protected from human sources of damage, since they are in areas reserved by commercial bird's-nest collecting interests.

The last group, similar to the latter in other respects, has been subjected to reef blasting by fishermen, with the result that only the larger massive corals have survived, and these show extensive signs of damage. In some areas where dynamiting took place several years ago (e.g. Raet and Lawa Islands), living coral has moved back in, with branching *Acropora* species predominating.

Overall, however, the reefs in the Chumphon coastal area are still in good condition. Recently, moreover, five islands around Chumphon have been designated 'reserved areas' by the Fisheries Department, mainly to protect their fringing reefs for tourism.

Koh Tao: Still in Chumphon Province, but far enough away from the coast (75 km) to be considered separately, this island lies in exceptionally clear water for the Gulf (some 15 meters of visibility), providing ideal conditions for coral growth. The high density of coral cover is made the more interesting for the presence of some species which are not found in other areas.

Because of heavy wave action during the northeast monsoon (December-March), coral cover is limited on the steep rocky shoreline of the eastern shore, and with some areas supporting only small corals of species such as *Pocillopora*, which can tolerate heavy waves, together with soft corals such as *Sinularia* and *Sarcophyton*. Some of the southeast reef flats, however, display 100 percent coral cover (much of it branching *Acropora*).

Until the late 1980s, very few people visited this island. Tourism has started to flourish, however, with bungalows being constructed on the island itself and ever-increasing tour-boat traffic, and this has had predictable consequences. Between 1985 and 1987, the ASEAN-Australia Marine Survey Project found extensive damage had already occurred, mostly due to the anchoring of tour boats. Since that time, the pace of destruction from tourism-related activities has accelerated (while the ravages of Typhoon Gay, in 1989, will be seen on the local reefs for some years to come).

One of the finest local dive sites is tiny Nang-Yuan Island, off the northwest shore of Koh Tao. Overall, in the two years between the surveys, the proportion of dead coral here rose from 0 percent to 18.5 percent. This has tentatively been attributed to both anchoring and to crown-of-thorns infestation. In Thian Bay, another prime site, over the same period coral cover decreased from 100 percent to 53.9 percent, with anchor damage again being the prime culprit.

Koh Tao has been slated for further tourist development, with policymakers convinced its potential as a money-earner has not been yet realized. Ironically, perhaps, this bodes ill for its chances of survival as a place worth visiting.

Ang Thong Group: Huddled together for mutual protection from wind and wave—this group of more than 40 islands in Surat Thani Province is not so spectacular for coral growth or diving, though there are a few good reefs. For one thing, in years past the local reefs suffered from extensive bottom-trawling. The area has since 1980 been designated a Marine National Park, which has provided some protection from human impact, though in heavily touristed areas there has still been significant damage from tour-boat anchoring and from collecting.

Natural pressures have also taken their toll, with sediment deposited by the Thapee-Phunduang River probably being the largest single factor. The density of living coral is not

Koh Phi Phi, with its spectacular limestone cliffs and secluded beaches, has enjoyed a reputation as being one of the world's most beautiful island groups. Unfortunately a variety of pressures, most of them related to commercial tourism, are degrading underwater habitats surrounding the islands despite the fact Koh Phi Phi has been designated a marine national park. **Above:** *Phi Phi Don becomes more popular every year.* **Below:** *This shot was taken on Phi Phi Le, the smaller and less developed of the two islands.*

Campbell's Bay, on Koh Similan (Island No.8) is a favorite anchorage for diveboats, yachts, and daytrippers. The Similan Islands, lying about 100 km northwest of Phuket Island, are frequently touted as being among the world's top ten sports diving destinations. A combination of interesting bottom topography, abundant fish life, and clear waters make this a fascinating area to explore. Dynamite fishermen were pushed out of the area by the advent of tourism. Mass tourism itself, however, brought pressures such as anchor damage and the dumping of rubbish, pressures potentially as devastating as dynamite. The designation of the area as a marine national park, plus conservation efforts by local dive shops and government agencies, has apparently turned the pattern of destruction.

very high in the areas worst affected, with the dominant species being those such as *Porites lutea* and *Goniopora djibouteonsis*, which have a relatively high tolerance for sediment. However, both high coral density and diversity can be found where there is clear water and strong water movement, such as off the western side of Sam Sao Island and the reef on the northern shore of Tai Plao Island, and the ASEAN-Australia Marine Survey Project report concluded the Ang Thong Group was overall the most 'natural' of the areas surveyed.

Koh Pha Ngan: Located far from shore and about 18 km north of Koh Samui, this island lies in clear water well removed from river-borne sediments, but muck stirred up by the constant traffic in tour boats has killed off much of the living coral cover on shallow reefs in certain areas. Local fishermen also follow the practice of tearing up living coral to use in camouflaging the large fish traps they set off the reef.

Marine survey reports have nevertheless found that corals off Pha Ngan are generally in good condition, though they warn of rapidly deteriorating reefs if conservation efforts are not instituted immediately.

The Islands South of Samui: These islands are separated from Samui by a 30-meter-deep channel, and the pattern of sea currents cuts off sediment from the Thapee River. Thus, underwater visibility is better than it is off either Samui or the Ang Thong group, and coral reefs are found on all the island here. These reefs tend to be smaller than those found off the former islands, but there is a higher density and diversity of living coral. The reef communities are also generally more complex, but with *Porites lutea* and *Acropora* predominating on many of the reefs.

Interestingly, and uniquely in Thai waters, the marine surveys have found some actual coral islands (as opposed to fringing reefs) in this area, though it is not yet certain how these arose. There are hopes in local scientific circles that these islands will be designated scientific sanctuaries, with all other uses prohibited.

No deterioration was found between surveys conducted in 1985 and 1987.

Koh Samui: Located 79 km from Surat Thani Town, this island has seen a massive increase in tourism during the 1980s, and development continues apace, with an airport and proliferating accommodations of every standard from primitive bungalows to five-star hotels.

The reefs on the western and northern sides of the island are adversely affected by sediments from the Thapee-Phumduang River, so visibility is poor and coral cover is not high. The east-coast reefs, especially in the southeast area, display both much higher diversity and density.

Dynamiting used to be the main human pressure on these reefs, but that practice has been largely discontinued. Anchor damage and collecting have now taken up where reef blasting has left off. There is also evidence that the western coastal reefs have been affected by sediment which has had its source in land development aimed mostly at the tourist trade.

Chonburi Province: In the inner Gulf, on the eastern side, one starts to find corals around Koh Sichang. Here, the reefs are in the 'coral community' stage, with no complexity of structure, and with *Porites lutea* the dominant species. Visibility is limited.

A bit farther down the Gulf, Pattaya and the islands which lie off this most commercially exploited of Thai seaside resorts have been popular dive destinations for decades. Unfortunately, anchoring, sewage and other waste disposal methods, collecting, and sediments from coastal development projects have reduced live coral cover and visibility to the point this area cannot be recommended for much more than scuba instruction, (although there are still a few interesting reefs off the so-called 'Outer Islands').

Still, even in waters such as those around Koh Larn, the largest of the islands off Pattaya, where the coral is effectively dead, recovery would be possible if the reefs were given a chance. Within a few years, according to local marine biologists, there would already be enough to make diving interesting (but during those first years extra care would be needed, since small corals are easy and tempting to pick up and take away).

On the plus side, there are excellent dive shops in Pattaya, and the area is easily accessible from Bangkok. Another advantage is that the local climate permits diving year-round.

An hour's drive farther south one comes to Sattaheep and Samae San, fishing villages from which boats can take you to reasonably unspoiled coral reefs (though dynamite fishing has been a serious problem in this region). Also worth noting are two sunken freighters: the *Bremen*, lying in 23 meters off Sattaheep, and the *Hardeep*, in 27 meters off Samae San. Both of these now constitute thriving artificial reefs in waters accessible to scuba divers. Visibility here can still be quite good, though sediments from coastal development, again, are gradually increasing the turbidity of the water.

Just south of the wreck of the *Hardeep*, there are fringing reefs which show an astonishing variety of underwater prospects, and a rich diversity of soft corals.

Trat Province: More than 40 islands, the largest of them Koh Chang, are to be found in this most easterly of the Gulf areas. This is one of the remotest areas, in terms of difficulty of access from Bangkok. So far, as well, it lacks facilities to handle tourism on any scale, though that is changing. Developers are already eyeing this relatively unspoiled area of islands and reefs as the natural successor to Phuket's 'success' story.

The best reefs tend to be those on the islands farthest offshore, though Koh Kradat has some good coral reefs, and parts of Koh Mak offer nice snorkeling as well. There are lots of gorgonians and soft corals, as well as a good variety of hard corals. Unfortunately, dynamite fishing has laid waste to many areas of what had been flourishing coral reef. Underwater visibility is quite variable, with runoff from freshwater streams, longshore development, and local gem mining operations sometimes increasing turbidity.

The Andaman Sea

The Andaman Sea exhibits a geography quite different from that of the Gulf of Thailand. The corals are found on a continental shelf which extends out from the mainland. This area is 150-200 meters at its deepest. It drops off to another level, which is about 300 meters at its deepest. This shelf in turn plunges to a third at 300-500 meters; and, finally, there is the sea floor, at more than 3000 meters. The scenery is frequently spectacular, with sheer limestone cliffs thrusting out of water that sometimes allows visibility to 45 meters.

As far south as Tarutao, the limestone is the result of carbonate deposition during the Ordovician and Cambrian periods. Through Krabi and Phangnga Bay the limestone is much younger, dating from the early Permian some 230 million years ago, and can be described as a fossil barrier reef, the remains of a series of coral reefs which stretched for thousands of kilometers as far as what is now southern China. During the same geological epoch, granitic intrusions forced their way up through the sedimentary strata in parallel north-south zones. One of these stretches from the Adang group of islands, just west of Koh Tarutao, all the way north to Phuket and beyond. Another links the Similan and the Surin Islands.

In the Similan Islands, jumbled piles of enormous granite boulders offer an interesting substrate for coral growth—the conditions for which are ideal, with a minimum prevailing sea temperature of about 28°C and a higher salinity than that found in the Gulf. You thus seem to find a greater variety of corals than you do on the Gulf side—more than 200 species belonging to 75 genera and subgenera have been recorded, as compared to 60 species in the Gulf—as well as a slightly richer variety of reef fish—106 species and 30 genera. (These figures are generally thought to be partial, however, and accurate numbers of species on both sides remain to be counted.)

The causes of reef degradation vary from site to site, but the major factors are crown-of-thorns infestations, storm damage, anchoring and other damage connected with recreational activities, blasting, and sedimentation from mining and coastal construction. The relative impact of human pressures is hard to assess definitively, since more research on longer-term variability of the coral reef ecosystem due to natural factors is needed. But, given the enormous and growing numbers of visitors to Phuket and environs, it is safe to say that commercial tourism is now the biggest single threat.

Koh Surin: Koh Surin is said to have the best-developed reefs, properly speaking, in Thai waters. The coral on these islands (two large, and two smaller) remains in generally good condition, though there is evidence of extensive dynamite blasting in some locations. As of this writing, tourism has yet to make much of an impact in the

Trat is so far little developed for tourism, though much is in the wind. This island group, on the whole, is lovely; but underwater visibility around the islands closest to the mainland is poor, while many of the fringing reefs farther out have been dynamited by fishermen. Raw sewage from villages and fishing boats such as those in the picture (taken on Koh Chang, Trat Province) already can have an adverse effect on the reefs. Imagine what full-scale commercial tourism might be capable of.

Located in the Gulf of Thailand about 80 km from Chumphon, on the mainland, and 55 km northwest of Koh Samui, Koh Tao lies in exceptionally clear waters for the Gulf (some 15 meters of visibility), providing ideal conditions for coral growth. Its ever-increasing popularity is bringing new pressures to bear on the coral, with tourist bungalows, shops, and tourboats proliferating rapidly.

The whale shark (Rhincodon typus), *the largest of the sharks, reaches lengths of 14 meters and more. Like the manta ray, the whale shark is harmless, feeding on plankton, squid, and small fish. Specimens are regularly spotted in the Gulf of Thailand, especially in the waters off Chumphon, and in the Andaman Sea around island groups such as the Similans and Surins. As with the mantas, they tend to appear in these areas more frequently during planktonic blooms. Also similarly, they are favorite playmates for sports divers. These creatures themselves appear to enjoy the relationship in some way, coming in again and again as though to court hitchhikers. (Richelieu Rock, Surins area in 9 meters.)*

The white-tipped reef shark (Triaenodon obesus) *grows to about 2 meters, and prefers depths of 30-40 meters, though it may be found in depths as shallow as 5 meters. It feeds mostly at night, generally staying close to the bottom. During the day, white-tipped sharks are often found resting in caves—the shark in this photo had at least five companions sharing the same shelter at the time. It has been a common belief that, since unlike the true fish they have no swim bladders, sharks must keep swimming or else sink and then die because they would be unable to breathe. But this species, at least, does rest.* T. obesus *lie on the bottom facing into the current with mouth open, passively allowing water to circulate around the gills. (Koh Surin in 9 meters.)*

area, it being until recently one of the frontiers for sports divers and well off the tour-boat path. Following the usual pattern, reef blasting has tended to disappear as tourism-related pressures such as anchor damage has increased. Recently, large cruise boats have put these islands on their itineraries, bringing hundreds of visitors at a time to the beaches and reefs.

At the same time, the Tourism Authority has installed mooring buoys, and some of what are often taken to be the most highly developed reefs in Thailand are still in good condition. The larger shellfish, however, are little in evidence, what with commercial collecting by the local Sea Gypsy (*chao le*) population.

Reefs extend 10-30 meters in depth, depending on the location of the islands, and coral cover varies depending on the degree of exposure to heavy wind and wave action of the coastlines.

Koh Similan: This group of ten small islands (if one extends honorary membership to Koh Bon, 25-30 km to the north, as part of the group) has developed a semi-legendary reputation among sports divers. About a 12-hour sail from Patong Beach on Phuket (90 km), or just a few hours by fast motor launch, the area offers lots of sites with breathtaking underwater scenes, whether in deep water or in water shallow enough in which to snorkel. Large fish are not so much in evidence as they used to be, due to excessive fishing (a similar phenomenon is already apparent in the Surins). The jumble of gigantic granite boulders is overgrown with a bewildering diversity of marine organisms, however, with the bottom off the west side of the archipelago providing submarine vistas which rank among the finest in the world. As with the Surin Islands, the reefs can extend from 10 to more than 30 meters in depth, and the underwater visibility is very good, occasionally ranging to 45 meters. Upwellings of cold, nutrient-rich ocean currents in late March and early April can temporarily impair visibility, but the resulting plankton blooms often attract deeper-water visitors such as manta rays and whale sharks. (It has also been theorized that upwellings such as these also help explain the special diversity and abundance of fish life in the Similans.)

The diving here is still extraordinary, but the pressure of its popularity is beginning to tell. There is a problem with rubbish being dumped on the reefs; and in some spots anchor damage has already created coral deserts, though daytrips to the Similans only began in 1987 (before that, there were just cruises of several days which often also took in other islands). By now there are about 35 tour operators on Patong Beach and in Phuket Town alone doing trips to the Similans. Large cruise boats are also bringing their hundreds of visitors at a time.

Efforts by local dive shops—mainly represented by the installation of moorings buoys and by attempts at public education—are having some success in turning the pattern of degradation.

Coastal Islands off Krabi: These islands are still for the most part in good condition, though the area is seeing a rapid growth in tourism and conservation measures are required now, before more significant damage is done. Periodic threats to begin a deep-sea port project in Krabi Province also alarm local environmentalists.

Phi Phi Islands: These two islands, among the most beautiful in the country, have had some of the best snorkeling and diving to be enjoyed anywhere. Already, however, reef blasting, crown-of-thorns predation, and insufficiently controlled tourism has degraded the coral to a considerable extent. In the surveys conducted in 1984-86, fewer than 45 percent of the reef areas studied were still in 'good' condition—that is to say, with more than 50 percent living coral cover.

Spectacular coral is still to be found a few kilometers north of Phi Phi Don, around Koh Mai Phai and on the reefs between Mai Phai and Koh Yeung.

The visibility, the diversity of fish, and the beauty of the surviving reefs still make this area extraordinary, however, and conservation measures might reverse the damage done.

Phuket Island: This largest of Thailand's islands, roughly the same area as Singapore, until the mid-1980s was renowned for the beauty of its beaches, the unspoiled nature of its resources on land and under the sea. By now it can boast the least interesting coral reefs in the Andaman region, with storms, crown-of-thorns predation and tin mining accounting for only some of the damage. The greater proportion of the destruction has been the result of unplanned and uncontrolled commercial tourism, including siltation from shoreline development, with anchor damage, collecting, untreated sewage disposal, and careless trampling also helping to do a very through job of destroying what were some of the loveliest reefs in the country. Reefs at North Patong Bay, Bang Tao Bay, and Chalong Bay are among those worst affected. (Hin Hmusang or 'Shark Point', as it is more commonly known, remains one popular spot for sports divers. Lying only a few kilometers off Phuket, it offers rocky reefs with a good variety of coral and fish and variable visibility. One of its biggest attractions for tourists are the eponymous leopard sharks, sizeable creatures so docile they allow themselves to be hugged. Unfortunately, in consequence, they have been molested to the point local observers fear they will soon abandon the area.)

Koh Racha: Racha Noi Island still has reefs in very good condition, with little damage reported, though the status of local fish populations has not yet been determined. Visibility is good, extending to 30 meters and more.

Racha Yai Island is in relatively good condition, with damage limited and found only in localized areas. The water quality is fair to good.

Tarutao National Park: The first marine national park in Thailand (established in 1974), these 51 islands are about as far off the beaten track as you will get, and they have so far been preserved from the worst effects of tourism. But partly because they are relatively remote, dynamite fishing in these waters has laid waste to considerable areas of coral, particularly in the 1970s. Pressures have more recently included serious infestations of crown-of-thorns. (It is not always easy to distinguish between old blasting and *Acanthaster* damage, since crown-of-thorns predation is often followed by erosion, and all that is left in either instance is coral rubble overgrown with algae.)

The waters around Tarutao Island itself, the second largest island in the Andaman Sea after Phuket, are shallow and rather turbid, with reefs along the northeast coast and in a bay on the west side of the island. The coral and the visibility is generally better around the islands just to the west of Koh Tarutao, especially Mu Koh Klang and the south-east and west coasts of both Koh Adang and Koh Rawi.

Finally, coral reefs, however relatively self-contained they may be, are part of a larger marine habitat (and, ultimately, a global ecosystem). Other coastal ecosystems are also being threatened by uninformed and careless exploitation. Two of these—the mangrove swamps and the seagrass beds—are important for both the well-being of Thailand's reefs and for a variety of commercial enterprises. A number of reef-dwellers, for one thing, depend on these other habitats for breeding grounds (if we can apply a generalization based on studies of coastal ecosystems in the Caribbean and North America). Not only do mangroves seem to play an essential role as nurseries in both reef and deep-water marine ecosystems, they also very effectively trap sediment coming down to the sea with freshwater runoff, thereby helping prevent corals from being smothered. (Just to add to the unfortunate chain of circumstances, the shrimp farms which replace the mangroves themselves produce a good deal of waste matter which winds up as sediment in the coastal seas.) Threats to the mangrove swamps—which are becoming critical—are indirectly threats to the coral reefs as well.

Which pressures are most threatening? The consequences of dynamite fishing and anchor damage are severe and readily apparent. Some of the effects of sedimentation and chemical pollution are also immediate and all too clearly visible. What the longer-term effects will be, however, is not so clear. But if we look at the consequences of pollution, collecting, anchor damage, and dynamite fishing together, it is obvious what will happen to an ecosystem so complex and still so largely mysterious as the coral reef. Without effective and immediate conservationist measures, environmental disaster is not far away.

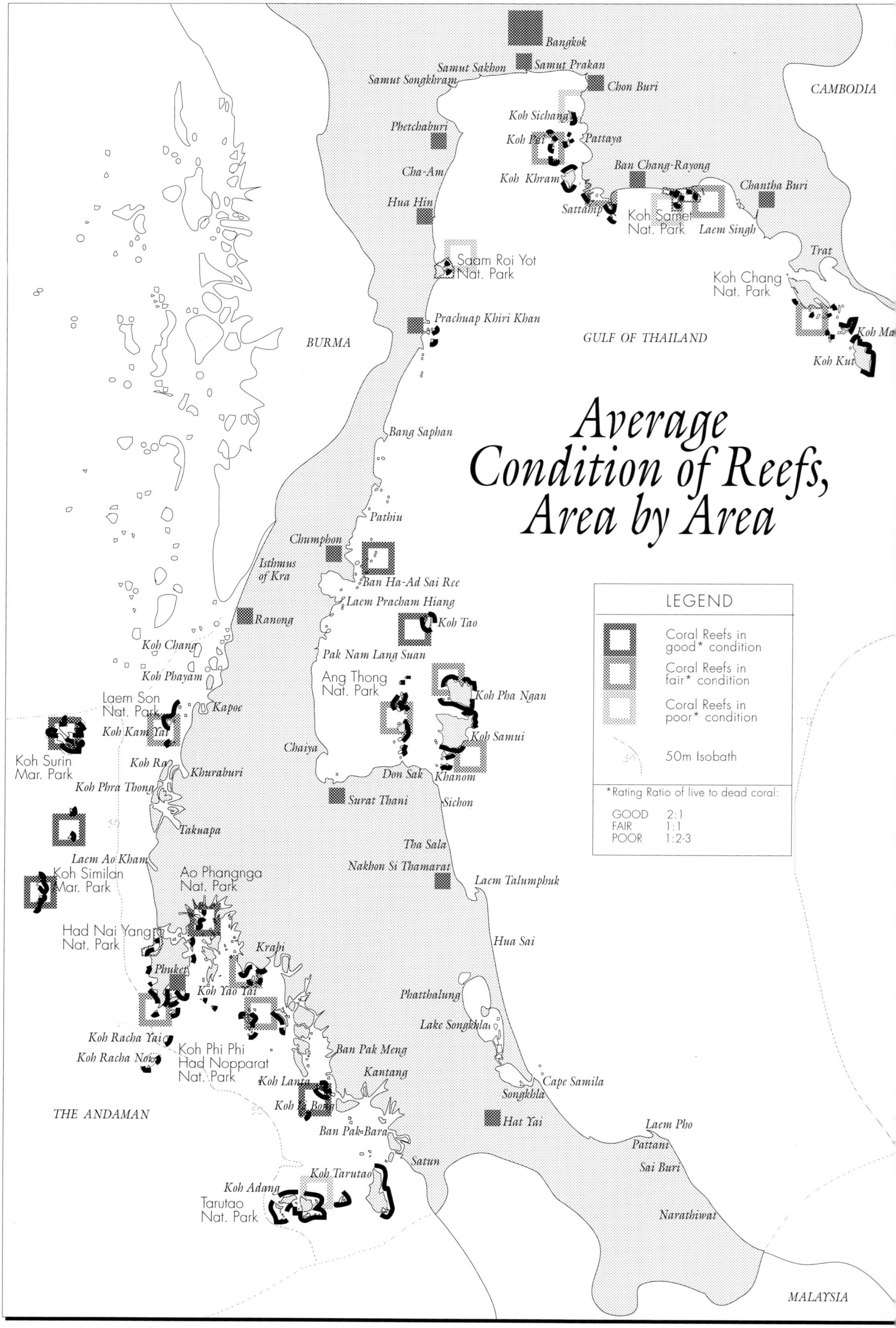

Average Condition of Reefs, Area by Area

PART THREE

THE CHALLENGE OF REEF CONSERVATION

Chapter VIII

What is Being Done?

The Coastal Resources Management Project (CRMP), under the auspices of both the National Environment Board of the Royal Thai Govern-ment and the Government of the U.S.A., has specified five main ways to limit reef damage:

a) the provision of information for both tourists and local residents;

b) measures to reduce anchor damage;

c) the reduction of siltation where practicable;

d) improved coordination among the concerned governmental agencies; and

e) more effective enforcement of conservation laws in nearshore waters.

It is much easier to enumerate these measures, however, than it is to find the public enthusiasm and funds necessary to effect them. In what follows, we will consider some of the things which are actually being done.

Reef surveys and other research:

There are several organizations now engaged in surveying the distribution and condition of Thailand's reefs. Such surveys try to monitor the following conditions: the proportion of live to dead coral; fish populations, including the presence of rare or uncommon species; evidence of disease or unusual predation (e.g. crown-of-thorns infestations); mechanical damage such as sediment smothering, anchor damage, blasting damage, and rubbish disposal; algal overgrowth; and, finally, the basic environmental parameters of turbidity, temperature, salinity, and available nutrients.

At popular tourist sites, the extent of anchor damage should be regularly monitored, as should water quality where reefs are threatened by siltation and poisonous wastes. Fisheries and other catch statistics, including shell and reef fish collection, should also be assembled and analyzed on a regular basis, with the results being made available to all concerned parties.

Agencies such as the Coastal Resources Management Project (CRMP), under the Thai National Environment Board, the ASEAN-Australian Marine Survey Project, the Phuket Marine Biological Center, a research institution under the Fisheries Department and, most recently, a new coastal resources management project (CORIN) at Prince of Songkhla University, are currently all conducting studies and surveys in both the Gulf of Thailand and the Andaman Sea.

Fisheries, for example, has declared the reefs essential breeding and feeding grounds for a variety of valuable marine life. Officials are now attempting to crack down on illegal fishing operations which are causing damage to this natural resource. The Department is also mapping and otherwise promoting coral reefs in the attempt to encourage more controlled and environmentally aware tourism development.

Public education and community projects:

An important purpose of these survey projects is an educational one: to inform the public, firstly, of the nature and distribution of this fast-dwindling natural resource and, secondly, to warn of the damage being done—of its true nature and extent.

On one hand, an attempt must be made to teach visitors an appreciation of marine natural resources—to make them more aware of what is at stake and how the reefs can be enjoyed without harming them. Measures already being adopted include expanded exhibitions at public aquariums and national parks, complete with informational brochures and audio-visual displays. Posting signs on beaches and in reef areas (together with effective enforcement of penalties for those who ignore regulations) are also part of any effective campaign.

Systematic education of tour operators and guides is essential. The CRMP has recommended a formal certification program for these occupational groups. Law enforcement officers must also be trained at the same time more effective and better law enforcement measures are adopted.

Local residents need information. A less exploitative attitude towards the reefs would surely follow, once it was understood that current practices are in the longer-term interest of neither fishermen, collectors, tourism employees, nor the public at large.

Some priorities are the monitoring, measuring, and clarification to local residents of the consequences of reef blasting, over-collection of fish and shell species together with the reef damage caused during collection operations, especially anchor damage. Such educational programs, it is argued, should be tailored to address a variety of audiences—for example the local Muslim and chao *le* (Sea Gypsy) communities—taking into account each group's particular traditions, interests, and needs.

In the opinion of the CRMP, educational programs and community involvement ideally should come together in a "community-based habitat management program" for reef areas (and, later, for mangrove swamps and seagrass beds).

Educational efforts by government agencies such as the Tourism Authority of Thailand (TAT), the Department of Fisheries, and the Royal Forestry Department (National Parks) would ideally be supplemented by contributions from tour companies and other commercial interests which benefit from the reefs.

In educating the local business community, the most direct appeal to their interests is an obvious one: Healthy coral reefs are good for tourism.

Formal training of tour operators, boat operators, and guides should be the responsibility of the TAT, community colleges, the Tour Guide Association of Thailand, the Fisheries Department, business associations, and National Parks. Law enforcement officers might be given the relevant training by the Marine Police as well as by the above organizations.

The CRMP recommends that school curricula adopt courses and class activities which would educate the country's youth. Money should also be spent on a media program: radio and television, in particular, speak to a large proportion of the population, and are used by the government as a matter of course to encourage public cooperation in other areas.

Another measure—one which both publicizes the issues and directly addresses some of the problems—is the organization of special community projects. This may be one of the surest routes to effective public education. Already, in Phuket and Pattaya, underwater cleanup campaigns have been organized which involved the enthusiastic participation of local diving associations, members of the general public, and even tourists. In Phuket, the Marine Biological Center has supervised a campaign to counter a crown-of-thorns starfish explosion which threatened decimation of local reefs.

Voluntary community activities have even been extended to mooring buoy installations. The Phuket pilot project, where mooring buoys have been installed in the attempt to discourage anchoring on the coral, was the first such effort. This has been followed by similar projects in the Gulf.

Public education would perhaps most importantly target business people, legislators, and policymakers at all levels. There is an urgent need for more carefully planned expansion of tourism in coastal waters. One initial step towards this goal would be a more coordinated approach to environmental problems in Thailand: policy planning and legislative control requires better coordination between the National Environment Board, the Department of Fisheries, the Royal Forestry Department, and other agencies.

It should also be noted that public education must be complemented by displays of good faith by government and business, including

Above: *A school of batfish (*Platax teira*) cruise at about 20 meters around the legs of an offshore oil platform in the Gulf of Thailand.* **Left:** *A bushy black antipatharian coral grows on the rusted hull of the* Hardeep, *a sunken freighter lying in 27 meters off Samae San, south of Pattaya. One might think that shipwrecks represent environmentally destructive litter on a truly massive scale. On the contrary, a wreck can provide an instant home to a wide variety of marine life. Eventually, in appropriate depths and with other conditions permitting, a thriving reef community is established.*

Above: *Although offshore oil platforms are more often associated in the public mind with marine pollution, they can in fact provide substrata for artificial reefs. Corals, algae, fish and a variety of other marine organisms establish thriving communities on and around the legs of the platform.* **Right:** *Rabbitfish (*Siganus guttatus*) graze on an oil platform, sampling from a lavish buffet of coral and algae. (Gulf of Thailand in 25 meters).* **Below:** *A number of* Abudefduf sexfasciatus, *coral reef fish more commonly known as sergeant majors, have found a new home around this platform, in about 28 meters.*

the allocation of funds to support the growth of more environmentally aware attitudes. For example, the problem of tour boat operators and tourists dumping rubbish on the reefs would be alleviated if proper waste disposal facilities were made available in appropriate areas.

Waste-water treatment facilities:

Once Thailand's 'premier beach resort', Pattaya Bay has been widely publicized as being so polluted by raw sewage that it is no longer safe to swim there. It took the loss of earnings caused by a marked drop in tourism to wake local business and government up to the problem, but steps being considered are still inadequate. Rather than merely treat the wastes so that the bacteria are killed, recycling plants are what is really needed. A few hotels have already undertaken their own recycling systems, though even they only are treating sewage, and neglect other waste water. The plan is for local hotels to cover part of the cost of building treatment plants.

It would be nice to think that the stink surrounding Pattaya's excesses would inform development plans elsewhere. Unfortunately, it sometimes seems that each area is determined to go through the same process of degradation of that very environment which supports it, effectively learning nothing from the experience of others.

Mooring facilities:

The first project the CRMP set up was one to establish mooring buoys in areas where anchor damage threatened to destroy the reefs. The pilot project was undertaken on Phuket for a variety of reasons. For one thing, the reefs there have great commercial value for both tourism and fisheries (aside from their intrinsic ecological and scientific value, of course). It was believed that reasonably inexpensive and effective means—mainly the installation of mooring buoys and public education—were available to halt and even reverse at least some types of destruction.

In Chumphon, many of the mooring buoys have disappeared, apparently because the anchoring points have broken, not being strong enough to secure the larger fishing boats which have used them. The buoys which remain are not used (which, it seems, is why they remain); they were installed too close to the rocks, and it isn't safe to tie a boat up to them when there is any sea at all.

The buoys have been installed both inside marine national parks (Phi Phi and Similan Islands) and outside (Koh Hae and Patong Beach). Plans call for the program to be expanded to embrace all popular tour and diving sites in the country. For the scheme to be maximally effective, it should be combined with legislation defining 'no-anchoring zones'. Ideally, the CRMP suggests, the whole program would be extended with the collaboration of the TAT, the Marine Police, tour companies and other businesses, and dive clubs. Volunteer divers could be used to anchor the buoys, while training to familiarize tour-boat operators with use of the buoys might eventually lead to their cooperation in regular surveillance and maintenance of the facilities. Given the attitude of most commercial interests to conservation up till now, this scenario, however desirable, may be somewhat utopian.

In the Andaman Sea area, so far, it has been just a few dive shops which have contributed most to these plans.

Artificial reefs:

An attempt may be made to preserve or to replace the essential habitat of the reef ecosystem by seeding new reefs. This no doubt faster than waiting for the natural process of accretion to take place, though the results, even where relatively successful, tend to be much less aesthetically pleasing.

In both the Phuket and Pattaya areas, old car tires have been put to use by wiring them together by the hundreds and sinking them at suitable sites. After a couple of years, however, these rubber reefs disintegrated. In any case, at least in the Pattaya area where the reefs were seeded, the turbidity of the waters probably would have meant this particular artificial reef wouldn't have 'taken'.

More successful artificial reefs, constructed of concrete, may be found in Ranong, with more to come in Krabi, Trang, and the Gulf. These reefs are not deliberate—they are in fact designed to serve other purposes, for example attracting a variety of fish species, and to keep trawlers from coming in too close and disturbing the waters around certain villages. The seeding of coral reefs has been merely a secondary effect.

One form of artificial reef much prized by sports divers is the sunken wreck. Fine examples are to be found near Sataheep and Samae San, a little way south of Pattaya. Although visibility is deteriorating in these waters as well, the steel-hulled wrecks of the *Bremen* and the *Hardeep* still teem with marine life. (In the wake of Typhoon Gay, 1989, there were moves to sink the wreck of the oil exploration drill ship *Seacrest* in waters where it might develop as a reef, but fears that it would be a navigational hazard led to its being towed to deeper water before it was scuttled.) Here are examples of littering on a gigantic scale which can actually serve the interests of marine life.

Similarly, offshore oil platforms quickly come to support thriving reef communities, with corals and algaes growing on the legs of the platforms, and many other marine organisms coming in to play their parts in the constitution of the new ecosystem.

Artificial reefs, though, would need hundreds of years to approach the complexity and sheer beauty of the natural ecosystems at risk, and ultimately they make a poor substitute for the conservation of what exists in nature now.

New legislation:

What is needed are reasonable controls aimed at sustainable development of this resource's potential. At present, two statutes regulate coral reef management in Thailand: the Fisheries Act of 1947 and the National Parks Act of 1961. Under these acts, 'marine protected areas' have been established throughout Thai waters. Since 1961, however, both circumstances and perceptions of the problems we face have changed drastically. In consequence, according to the government, the effectiveness of existing legislation is currently under review.

The National Park Act merely states that it is against the law to collect for commercial gain or to degrade any natural resource. The legislation was drafted primarily with hunting on land in mind, however, and is not generally understood to apply to marine resources.

As it stands, the collection and export of coral is prohibited. The use of poisons and explosives are against the law, as are trawling and push-net operations within 3000 meters of shore. Even assuming that these laws may be better enforced in future, much more than that is required. New laws are needed to regulate the treatment and disposal of wastes. The CRMP, for one, has suggested other changes including the expansion of tourism to waters farther offshore, which would have the effect of diluting the damage, if nothing else (though it might at the same time reduce the number of unspoiled reefs in Thailand). The reform of marine conservation legislation should also include the following: size, catch, and season limits for collected specimens; recognition that entire productive habitats (for example coral reefs and mangrove swamps) are of more central concern in fisheries than is single species regulation and management; prohibitions on the collection of rare species, especially the export of aquarium fish and threatened shell species; the monitoring of the export of all reef animals; and, very importantly, radically improved enforcement of all laws.

One thing that should be added to this list of requirements is improved coordination among the concerned agencies—policies governing the use of reefs in one area or another, including development planning; monitoring of reef use and conditions; and enforcement all need to be brought under a single scheme if a conservation campaign is to prove effective.

In 1990, the Thai cabinet approved what is termed the country's first 'conservation master plan'. This plan, proposed by the Ministry of Science, Technology and Energy, called for the setting up of a new division of the National Environment Board—the 'Environment, Nature and Arts Conservation Division'. At the grassroots level, according to the proposal, local councils are to play a 'key role' in the conservation and management of local natural resources, while the NEB will make policy recommendations and evaluate measures undertaken.

Lest this sound like just one more level of duplicated effort and proforma concern—one more accretion on a large bureaucratic structure that "resembles a human about as much as a reef resembles a coral polyp" (John Perry

Barlow, talking about bureaucracies in *Harper's*, March 1990)—the proposal also suggested the government bill should clearly define the responsibilities of all agencies involved in the conservation effort.

Increased fines for offenses are in the works, while p*ossession* of corals is to be deemed an offense as well as selling or exporting it. This legislation has already been held up for years, however, by the problem of determining specifically which organisms are to be prohibited as well as the problem of how to enforce the prohibition as it might apply to small quantities held for non-commercial use—is o*ne* piece of coral, for example, to be considered an offense?

More effective enforcement of existing laws:

Successful enforcement of laws, both existing and proposed, presuppose the improved training of enforcement officers, more personnel, and more equipment, including boats. All of this, of course, will cost money.

Enforcement of the national park laws is notoriously difficult even in Thailand's forests. Enforcement in the marine national parks faces all those problems plus the disadvantage that anyone can enter a marine national park without challenge—effective access is not limited to roads, as it is in the forests. On top of that, there is a serious lack of suitable patrol and pursuit craft.

One measure that is being considered in the short term is increasing the size of the fines.

Area Management Plans (Zoning)

Reefs should be zoned according to planned use. Especially sensitive areas should enjoy a total ban on anchoring. At the same time, some waters could support fishing and tourism, while others might be designated off-limits for any other than scientific uses. Just such a system has been established, with considerable success, on the Great Barrier Reef in Australia.

And perhaps zoning regulations could help minimize conflicts among users of coral reefs. For example, the interests of land developers and sea developers could be coordinated, as could those of fishing operations and tourism.

One proposal which has been aired recently is related to the idea of zoning resource use. This notion is that the TAT should be authorized to close areas such as Koh Samui and Koh Similan to tourism, from time to time, to allow the environment to recover. Just as we rotate crops to rest the soil and reduce pest populations, coral reefs in certain areas could be periodically put off-limits to give them the chance to recover, for example, from the pressures even enlightened tourism inevitably brings to bear.

Given the economic interests involved, however, it is probably safe to say this is one measure, however effective it might be, which will never be taken.

In any case, it may be suggested that laws and improvement enforcement procedures are not enough in themselves to ensure the sustainable use of the coral reefs. Of all the other measures outlined above, informed local involvement in resource management is perhaps the most important. Local interests must come to appreciate how the prosperity and well-being of the community is linked to economic growth which recognizes the need to reconcile potential conflict in the multiple use of the reef, while maintaining the health of the reef as a whole.

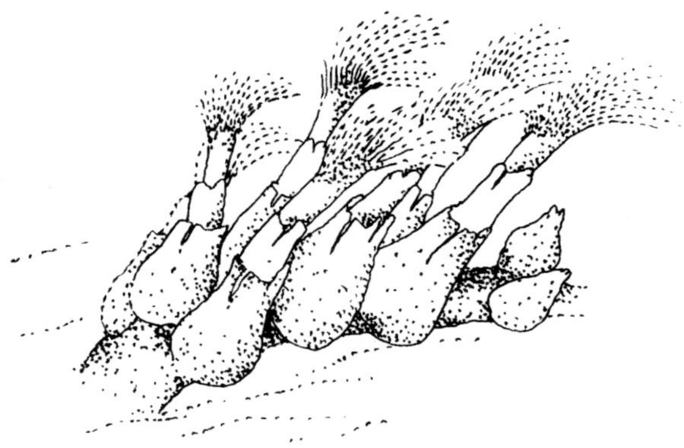

PART FOUR

PROGNOSIS

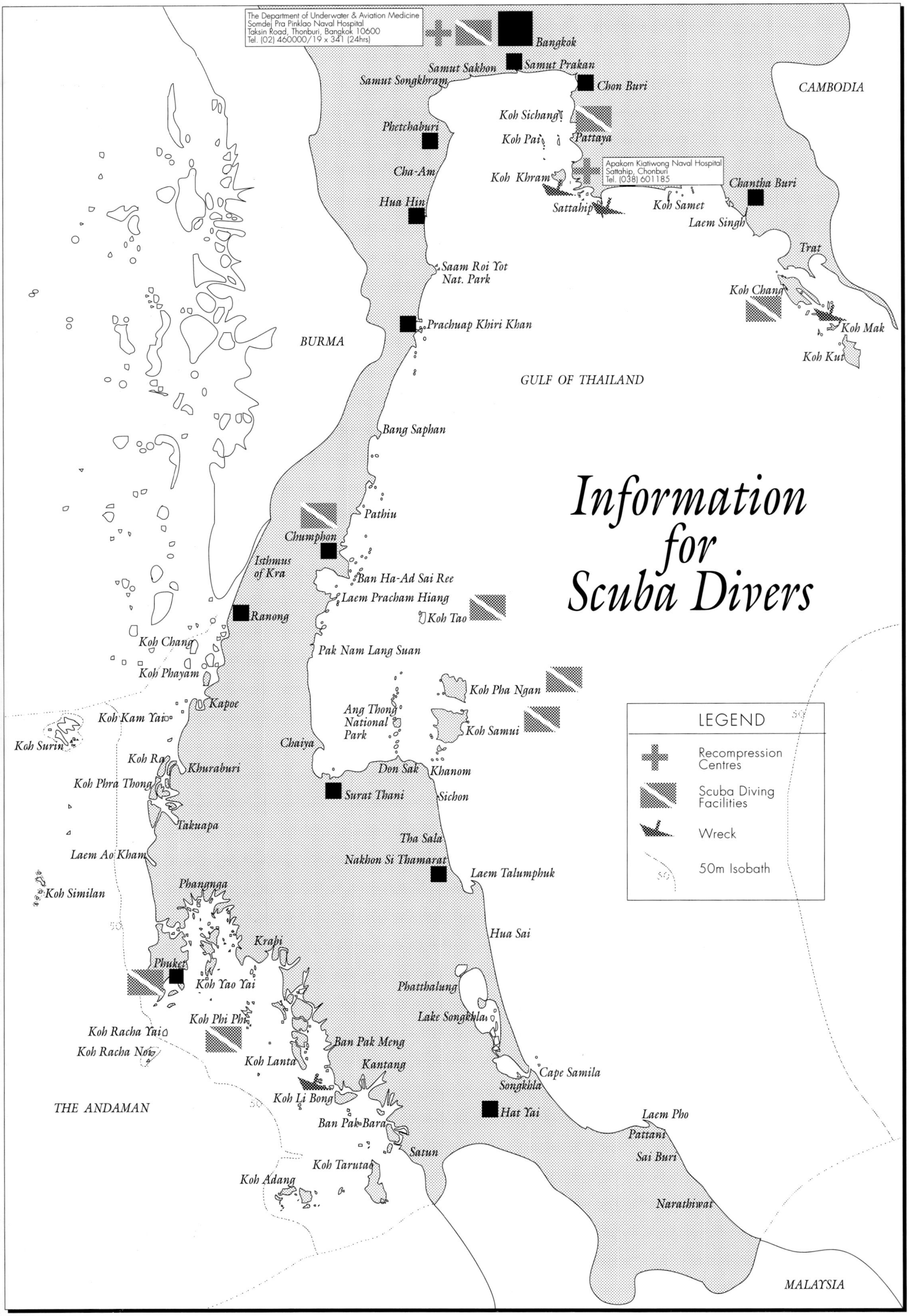
Information for Scuba Divers
The Department of Underwater & Aviation Medicine
Somdej Pra Pinklao Naval Hospital
Taksin Road, Thonburi, Bangkok 10600
Tel. (02) 460000/19 x 341 (24hrs)
Bangkok
Samut Sakhon
Samut Prakan
Samut Songkhram
Chon Buri
CAMBODIA
Koh Sichang
Phetchaburi
Koh Pai
Pattaya
Apakorn Kiatiwong Naval Hospital
Sattahip, Chonburi
Tel. (038) 601185
Cha-Am
Koh Khram
Chantha Buri
Hua Hin
Sattahip
Koh Samet
Laem Singh
Trat
Saam Roi Yot
Nat. Park
Koh Chang
Prachuap Khiri Khan
BURMA
Koh Mak
Koh Kut
GULF OF THAILAND
Bang Saphan
Pathiu
Chumphon
Isthmus
of Kra
Ban Ha-Ad Sai Ree
Laem Pracham Hiang
Ranong
Koh Tao
Koh Chang
Pak Nam Lang Suan
Koh Phayam
Koh Pha Ngan
Kapoe
Ang Thong
National
Park
Koh Kam Yai
Koh Samui
Koh Surin
Chaiya
Koh Ra
Khuraburi
Don Sak
Khanom
Koh Phra Thong
Surat Thani
Sichon
Takuapa
Tha Sala
Laem Ao Kham
Nakhon Si Thamarat
Laem Talumphuk
Koh Similan
Phangnga
Hua Sai
Krabi
Phuket
Koh Yao Yai
Phatthalung
Lake Songkhla
Koh Phi Phi
Koh Racha Yai
Koh Racha Noi
Ban Pak Meng
Koh Lanta
Kantang
Cape Samila
Songkhla
Koh Li Bong
THE ANDAMAN
Hat Yai
Laem Pho
Ban Pak Bara
Pattani
Satun
Sai Buri
Koh Tarutao
Koh Adang
Narathiwat
MALAYSIA
LEGEND
Recompression Centres
Scuba Diving Facilities
Wreck
50m Isobath

Chapter IX

What Has to Be Done?

Hundreds of millions of them living piled one atop the other, cheek by jowl in little cemented cells. As one individual dies, another builds on top of his predecessor until the colony is hundreds of meters thick... At the rate humanity is expanding, we might be describing a city of the not too distant future. But of course we are not; we are referring to the coral polyps and the reefs which are their cities.

These organisms ought to be quite comfortable with their living arrangements, having evolved this accommodation over many millions of years. But now they are threatened by another species, a creature which has exploded out of control, proliferating wildly, spreading destruction and poisonous wastes wherever it goes. Humankind is killing the reefs. And more than the coral polyps are threatened. There are also the rich marine ecologies which depend upon the reef, ecologies incredibly diverse, unbelievably beautiful, endlessly fascinating and potentially useful to pure science and practical medicine alike.

Three hundred years ago, when many of the massive coral colonies in existence today were just getting started, the human population of the entire world was only 470,000,000. Less than 200 years ago, in A.D. 1800, it was 919,000,000—less than the current population of China alone. Now the Earth as a whole is being asked to support more than five times that number of people, and recently the United Nations has revised upward to 6,251,000,000 its projection of the world's population by the end of this century, with three people being born every second and a population equivalent to that of Thailand's being added to the world every year.

This generation of humanity, furthermore, behaves as though it will be the last (and so it may be, say the pessimists). The natural resources of this planet are finite. We treat them as though they were infinite. Infinitely expendable. People have always done this, of course. If you needed more wood for the fire, you cut it. If you didn't really *need* more wood, but just wanted to have a lovely big bonfire, no problem—there was lots of wood. Then we were few. We could cut all the trees, collect all the cowry shells we wanted. There was always more and plenty.

But now we are many, and we are growing fast. And we live as though there will be no children. As though there will be no tomorrow. It is still possible for us to view the wonders of the tropical reef, although every year we have to travel farther and dive deeper to find undamaged coral, intact marine ecologies, and fish which have not yet learned to shy away from human intruders. The next generation may be able to enjoy these experiences only vicariously, through films and books which show things the way they used to be.

We periodically destroy cities of our own. These, however, can be reconstructed in a matter of a few years. Dresden, Hiroshima, Nagasaki are all once again thriving centers. On the other hand, a single reef of the sort which attracts tourists and collectors to places like the Similans or Koh Phi Phi, once destroyed, can take much longer to rebuild itself, and then only assuming it is given the opportunity to do so.

From the beginnings of their history people have fished on the reefs, made jewelry from coral, and used beautiful shells as currency. But in the past our numbers were few and our technology limited. We were part of an ecological system which tended to operate for the general benefit of its member species. There was death and destruction, but it was in balance with rebirth and renewal. Now, though modern man is no doubt still part of nature, we are no longer in balance with our environment.

Above: *Sea fans, leather corals,* Dendronephthya *soft corals, acropora corals, golden damselfish and many species of marine life flourish at about 24 meters off Koh Payu, Similans.* **Below:** *Koh Similan, Fantasy Reef, in 15 meters.*

Added to our sheer numbers and the basic drive of this mass of humanity to consume what is necessary for life and comfort, we now have the noxious excretions of our industrial economies. Men have polluted the world's seas with an estimated 20 billion tons of garbage, with everything from plastic bags to insecticides to drums of radioactive waste. Penguins in Antarctica have for years been contaminated by DDT. Nowhere have the seas been spared oil pollution. What we don't consume, we poison. We smother. We irradiate. Our love of novelty and adventure is followed hard upon by a magpie acquisitiveness and a genius for commercial exploitation which has led us everywhere, our technology making virtually every place on the globe vulnerable to our predations, whether on land or under the sea.

But resources such as the reefs belong equally to future generations. We have no right to deny our descendents what is theirs, while we have every responsibility to preserve it. As we have seen in the last chapter, the practical measures required to save the reefs are clear. And the economic means to effect these measures would be available, were public priorities other than they are. What is not clear is whether the political will can be generated to see that the appropriate measures are taken.

We have looked at what ostensibly is being done to protect the coral reefs. But how much is talk, and how much is action? How much is simply preaching to the converted, and how much are the general public and, more importantly, government and business interests actually being reached? (To what extent are government and business interests currently distinguishable one from the other?) To what extent are business interests pursuing only short-term interests, and to what extent is it realistic to expect government and business to take a longer-term perspective on the exploitation of natural resources? Given the scale of the national and international campaign to save the rainforests, and the negligible effect this has had on destructive tree-cutting practices by local interests both here and in Burma (with Cambodia being next in line for wholesale clearing of old-growth forest), our conclusions must be pessimistic. With all of the exposure given to the negative consequences of unsustainable exploitation of land resources, how much harder will it be to muster interest and support for the conservation of resources like the coral reef, which seem more remote still from the average person's sphere of concern?

As of 1989, the National Oceanic and Atmospheric Administration in the U.S.A. reported that fully one-third of America's coastal waters were too polluted for commercial shellfishing. At an environmental conference in 1990, Britain was criticized by Scandinavian countries for its 'unwillingness' to halt the dumping of sewage and industrial waste into the North Sea, already one of the world's most polluted bodies of water. (The Dutch Ministry of Transport has claimed that 100,000,000 tons of sewage sludge and 40,000 tons of heavy metals are dumped each year into the North Sea.) With all their advantages—with all the relative material wealth of the so-called developed countries, with literacy rates approaching 100 percent, with a free press, with long traditions of public debate and criticism of policymakers—these peoples are allowing vital natural resources to be ruined by interests which calculate advantage only in the short term. What hope then for the developing nations—particularly those rushing headlong towards 'Newly Industrialized Country' status?

The fundamental obstacles to effective conservation policies are cultural and philosophical in nature, and they are worldwide.

Humankind is part of nature, yet we have through our obsession with control and through our technology set ourselves as over against the rest of nature. Rather than seeing the environment in a real sense as part of us, as something which is the condition of our success, we view it as an obdurate and sometimes dangerous opponent from which we must wrest control and wring maximum material satisfaction.

These ideas are relatively new to history. They have arisen together with a positivistic science and technology which has become the final arbiter of truth and right, grounding its legitimacy in the patent success it has had in manipulating our material environment—for there is no doubt we have achieved a degree of control never before approached in history, bringing to many in the world an unprecedented material security and comfort. At the same time, however, we have been to some extent spiritually and intellectually impoverished by the mechanistic, overly materialistic world-view which has accompanied this spectacular success in other dimensions of our existence. Living in our cities and on our large-scale mechanized farms, we are cut off from the sense of belonging to a natural ecosystem. The only costs we measure are those of human effort, while costs in terms of other living matter—the coral reefs and forests, for example—and in terms of the degradation of the natural environment tend to be ignored. We have lost our sense of the sacred in nature.

> This we know: the earth does not belong to man; man belongs to the earth... Whatever befalls the earth befalls the sons of the earth. Man did not weave the web of life; he is merely a strand in it. Whatever he does to the web, he does to himself.
>
> Chief Seattle

In 1854, the Indian chief of a territory which is now the State of Washington sent a letter to the president of the United States, and in it he expressed sentiments which are the more poignant for the circumstances under which they were uttered—the final conquest of a native people's lands by the ruthless ideology of development and commercial progress.

The conception of humankind as an aspect of a single whole which includes the rest of nature has also been traditionally part of Buddhist thinking. Within Thailand's official Theravada Buddhist Sangha, the dharma, or prescribed discipline, has come historically to be practiced largely for the sake of one's own enlightenment as an individual, while social concerns have tended to be viewed as a distraction from that primary goal. Nevertheless, a growing number of Buddhist monks and teachers have taken up the cause of conservation, seeing it as part of a larger task of public education. More often of late, seminars and talks are being conducted on the challenge of using basic Buddhist teachings to educate Thais, who are 95 percent Buddhist, to have a greater environmental awareness. Essentially the message is that one which Chief Seattle had for the government of the United States almost 150 years ago: Man is part of nature; he can't exist outside of it. To the extent he abuses nature, he abuses himself.

In answer to the currently popular 'clean technology' answer to problems of the environment, these Buddhist teachers would suggest instead that greed is the basic issue, and all the clean technology one can hope for will have finally only a palliative effect. (One dimension of the problem is this: can we credibly expect everyone on earth eventually to share the standard of living now enjoyed, for example, by the Americans or Japanese—especially when that affluence is taken together with the continually rising level of expectations built into the ideology of progress and material development, and in view of the fact that the world's total population is fast approaching the six billion mark? Are the natural resources at hand infinite, then?) The problem of our relation to the environment must be addressed on a more basic level by means of turning consumer behavior towards a simpler style of life, by challenging a way of life which has become a religion of Consumerism, justifying all in terms of material gain and seeing nature as something outside humanity, as something which is there just to be carelessly exploited. We could take a lesson from the fine economy with which the reef itself, in its undisturbed natural environment, manages its resources—conserving and recycling essential elements within a balanced and sustainable system.

It may be suggested that rampant greed, and not our exploding population, is the basic issue. Attitudes which would lead us not only to consume less and share more, but also to readopt a sense of the sacred towards the world around us would do more to save the environment than any amount of technological problem-solving. It is this sort of approach, perhaps, which has inspired the 'Magic Eyes' campaign in Bangkok, which has not exhorted people to clean up the river or to stop dumping rubbish in it—rather, it has simply asked that people love the river. The implicit understanding is that what we love we will automatically also care for.

Unfortunately, it is much harder to aim credibly at creating a public love for the coral reefs. To begin with, the reefs don't share that rich history of human association

This massive Porites *coral head, in 6 meters off Koh Hu Yong, Similans, was shallow enough to get clipped by a boat motor. (Koh Hu Yong, Similans, 5-7 meters.)* **Inset:** *The fallen piece has also damaged the coral colony beneath it.*

enjoyed by the 'River of Kings'; nor do they present themselves daily for the people's inspection in the way the river does as it runs through the heart of Bangkok. When children can't swim for fear of illness, or when boat taxis come across patches of 'dead' water—black, anaerobic water in which nothing can live—or when there are no more fish to catch, people have to take notice. Not so with the destruction of the reefs.

Despite increasing protest by concerned environmentalists in the popular media, the reefs are still largely outside the public consciousness. They are far away and out of sight, and therefore irrelevant. Even in areas where many people depend on them for their livelihood, there exists a tacit belief that, since they've always been there, and generations of people have used them, they'll always be there. This licences an unreflective, irresponsible impulse to take as much profit in the short term as possible, and let the future look after itself.

These attitudes, as we have argued, require educational measures which address people at the level of basic aims and values. By history and tradition, Buddhist monks in this country have been more associated with the forests, and have had little to say directly about Thailand's marine natural resources. Their essential message regarding the forests applies equally to the coral reefs, however, as well as to the mangrove swamps and seagrass beds which are also currently being threatened with destruction. Unless modern societies come to reinterpret their relationship to the natural environment, grounds for optimism concerning the survival of our coastal resources, regrettably, are slight.

Let us imagine that we do engineer the demise of virtually every coral reef on Earth. The coastal waters of countries like Thailand are left a marine desert. So big deal. So what? Such indifference may be appropriate, at least from certain lofty cosmic perspectives.

Try it, for example, from this point of view. There has been life on Earth for at least three billion years. The first large-scale reefs were established two billion years ago. The earliest reef-builders were not corals, mind you; they were primitive mobile organisms called archaeochathids, and at one time they were the dominant life-form on the planet, establishing all through the warm shallow seas those layered mats the fossilized remains of which we describe as stromatolites.

For a very long time—until 600 million years ago—these organisms flourished, pretty much lords of all they surveyed, for what that was worth. But the best-laid plans of mice and archaeochathids... After all those eons, a so-far-unidentified but surely cataclysmic event or series of events wiped them out.

By a time roughly 100 million years later, a more complex reef community had already evolved in their place. The main builders, this time around, were various species of calcareous sponge, though bryozoans, colonial algae, and the first two known orders of coral also played roles in the process. Thereafter, for 100 million years or more, the reefs did very well, growing in complexity and incidentally welcoming the signal success of the fishes, which had appeared earlier, during the Silurian Period, and which now, in the Devonian, were diversifying and filling all sorts of likely niches.

Then, with a fine disregard for all these accomplishments, another unknown disaster on an earth-shattering scale killed off the reefs together with most other creatures on the globe.

Evolution's natural exuberance is not easily dampened, it seems, because once again before too long new life arose on all sides. The reefs were evidently one experiment worth trying again; and once more the reef community was a novel one, this time with the feather stars becoming the dominant species, in association with bryozoans and others.

Almost predictably by now, had there been anything in the neighborhood capable of prediction, disaster struck again, this time removing from the course of natural history more than 50 percent of all the marine families in existence.

The process of recovery began once more, only to be brought to a rude near-termination by the same circumstances which ended the rule of the dinosaurs on land some 60 million years ago. (Some evidence has it that an asteroid or asteroids hit the planet with the force of millions of hydrogen bombs, resulting among other things in the equivalent of a nuclear winter).

Today, however, the coral reefs have reestablished themselves for at least a fourth time. In one form or another, by now, they have now persisted for 450 million years. (The original corals became extinct 200 million years ago though their reefs, it is believed, were quite similar to those constructed by today's scleractinian corals.) And since the last great extinction, as we have seen in earlier chapters, the reefs have survived everything from periods of devastating vulcanism to widely fluctuating sea levels to association with human beings.

So perhaps nothing is new, after all. The environmental catastrophe we threaten to bring about would be just one more episode in a long history of similar planet-wide mass extinctions and possibly not even a very impressive one, as these things go.

One could argue that there are interesting differences, though. Human beings are the first living organisms in our natural history which are capable of utterly destroying their own environment. (That is in fact true only if we overlook those early single-celled organisms which probably were first responsible for creating an oxygen-rich atmosphere. Oxygen was for them a toxic byproduct of their own metabolism, and so far did they pollute their environment that most species died out, making room for the evolution of creatures such as ourselves which actually depend on that oxygen, their waste, for life.) Another feature, this one certainly a novel one: we will also be conscious of what we have done and what we will lose.

Corals have thrived for 450 million years, despite some iffy episodes. We humans, on the other hand, have been around for just 3-4 million years. And consider this: of all species which have ever lived, it is estimated, fully 99.9 percent are now extinct. Many of these creatures, furthermore, did not suffer our disadvantages—i.e. this insatiable drive to consume combined with a cleverness sufficient to pander to this lust on a massive scale.

What we do to our environment we do to ourselves.

APPENDIX

THUMBNAIL PROFILES

PLANTS (algae and sea grasses)

The marine plants have been around for more than two billion years. They are the primary producers, forming the basis of a food-chain which allows the coral reef ecosystem as a whole to persist.

Sea grasses (phylum *Tracheophyta*, family *Hydrocharitaceae*) and algae (phylum *Chlorophyta*) often constitute a distinct zone on the sandy or muddy reef flat, with many fish and invertebrate species depending on this habitat at some point in their life cycles.

Sunlight, carbon dioxide, and water is photosynthesized to produce the sugars and other complex hydrocarbons required by the reef community. At the same time, large quantities of nitrogen and other elements essential to the growth of the nearby living reef are fixed. (Interdependence is evident at every level—the seagrass-algae community is in its turn protected from wave action by the reef.)

Surprisingly enough, few reef animals feed directly on the living sea grasses — with sea urchins, sea turtles, and dugongs being among the exceptions — and the most generous contributions to the food chain are made through fungal and bacterial decomposition of dead plant tissue.

The algae represent the other main food source for a wide variety of reef animals, with everything from crustaceans to fish to urchins to mollusks to, again, the sea turtles and dugongs feeding directly on the algae. As we have already seen, symbiotic algae (zooxanthellae) living in association with the hard corals and other organisms also play a vital role in the life of the coral reef. Beyond that, coralline algae makes an important contribution to the cementing of the reef substrate, flourishing particularly in areas of high wave action and thus making the reef more resistant to erosion.

An interesting footnote: one of the waste products from the photosynthetic activity of the ocean's algae is believed to have made life as we know it on Earth possible—most of the oxygen in our atmosphere has had its source here.

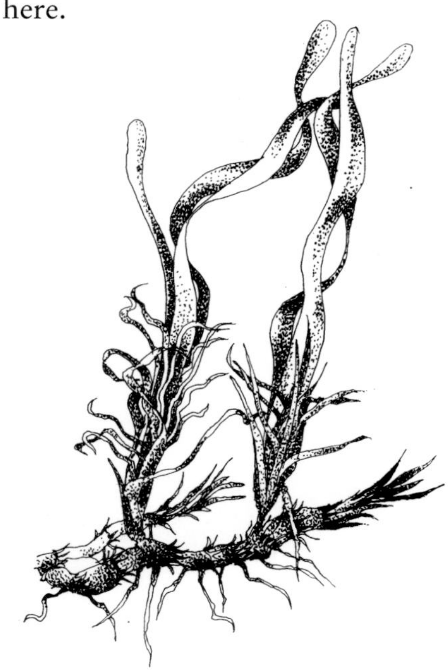

Less celebrated a member of the reef community than the hard corals, encrusting coralline algae is nevertheless an important reef builder. (Koh Payu, Similans, in 6 meters.)

Encrusting coralline algae flourish at the seaward edge of the reef and in other areas of high wave action, thus making the reef more resistant to erosion. (Koh Surin in 15-18 meters.)

This organism remains unidentified, though it is probably some sort of algae. (Koh Similan in 14 meters.)

With its jointed segments, Halimeda opuntia *bears some resemblance to a cactus. In fact, it is a calcareous algae and, as such, contributes calcium carbonate to the reef structure. (Koh Hu Yong, Similans, in 9 meters.)*

Chlorodesmis fastigiata, or 'turtle grass', is really a type of algae. A vital link in the food chain, this plant helps feed everything from turtles to sea urchins to bivalves and gastropods. (Koh Similan in 14-16 meters.)

This vivid blue sponge was photographed off Hin Pae, Chumphon, in 14 meters.

Brown algae such as this Padina tenuis *typically grow on rocks in very shallow water. This one, however, was found in 15 meters off Koh Ba Ngu, Similans. Some* Padina *are actually colored grey, due to their lime content.*

SPONGES (Phylum *Porifera*).

Probably the first and still the simplest multi-celled animals, having no specialized tissues or organs, sponges go back 650 million years to the Precambrian period.

Modern sponges are quite similar to their distant ancestors, but their relative status in the reef community has changed radically. During the Precambrian, sponges with calcareous skeletons, the stomatoporoids, formed much of the substrate of the world's reefs, with ancestors of the modern corals making a relatively minor contribution. But by the Devonian period, 420 million years ago, hard corals had usurped the role of primary reef-builder. (It is theorized that the sponges lacked symbionts which could aid in calcareous skeleton formation, and hence lost their primacy to the faster-growing corals, which enjoyed the collaboration of the zooxanthellae.)

Some species of sponge still contribute to reef-building, however paradoxically, in the process of boring into coral. Although about 10 percent of the limestone skeleton is destroyed through chemical dissolution, at the same time both coarse and fine sand is produced; this settles in holes and crevices and is subsequently cemented by coralline algae, eventually adding to the real substrate—the reef rock—and thereby lending further protection from really destructive forces such as typhoons.

The tube systems of the sponges also provide (non-mutualistic) habitats for annelid worms, crustaceans, brittle stars, and, sometimes, algae. Bryozoans and barnacles may attach themselves to the exteriors of sponges.

Today there are almost 5000 species of sponge, coming in a variety of colors and shapes and ranging in size from one centimeter to a meter and a half and more. Despite defensive adaptations such as mineralized sclerites, noxious chemicals and tough fibers, they are often made a meal of by vertebrates such as angelfish and filefish. Some nudibranchs, starfish, sea urchins, and cowries also favor sponges, immune as they are and hence free to ignore the bright colors which often warn of toxins.

Many sponges are 'chimney feeders'; that is to say they are funnel- or tube- or cask-shaped, and passing currents create updrafts which draw water in through the thousands of pores connected to interior systems of canals. These canals are lined with hair-like cells which beat rapidly to create internal currents which draw organic detritus and plankton-bearing water inside, trapping foodstuff and absorbing oxygen. Barrel sponges (*Petrosia*) can reach a height of almost two meters.

Sponges can reproduce from tiny pieces. Even where they are decomposed into separate cells, these cells will reaggregate and go on to constitute a new individual of the species.

Most sponges are hermaphroditic. The egg is fertilized and develops into a flagellated larva inside the parent body. This larva does not travel far before settling down to begin feeding and growing. Growth can be very rapid; under ideal conditions, sponges have been observed to double their weight in two days.

In species where there is only one sex cell, animals with eggs must wait for sperm to drift in on the current. Sponges may also bud asexually; and some produce spores called gemmules, which are collections of cells encased in a tough wall which can survive extremes of temperature or dehydration and still develop later into new animals. All these reproductive strategies have together helped to ensure the notable success of the phylum.

Sponges have been used by humans mainly as an aid to bathing, at least until recently. Lately, however, these lowly creatures are offering promise of much greater benefits. Sponges would be at a disadvantage, compared to animals which have developed spines or shells, except that like many other soft-bodied marine invertebrates they have evolved an impressive range of chemical defenses. These are generally toxic for predator fish, but they have been found to have at the same time a wide range of pharmaceutical properties for human beings. Medical research has already derived Acyclovir, which is used to treat herpes and encephalitis, from a Caribbean sponge. Another sponge (*Luffariella variabilis*), this one from the South Pacific, has yielded manoalides, a class of compounds which are undergoing clinical tests for use in the treatment of everything from arthritis and muscular dystrophy to psoriasis. Other substances found in a variety of sponges are being tested for their effectiveness against parasitic worms, the AIDS virus, and more.

Scientists have just begun to explore the possibilities of what appears to a pharmaceutical treasure trove. With an estimated 73 percent of the sponge, echinoderm, ascidian, and coelenterate species already tested being poisonous to fish, the coral reef represents an incredibly rich medical research laboratory, with a world of new chemical substances awaiting discovery.

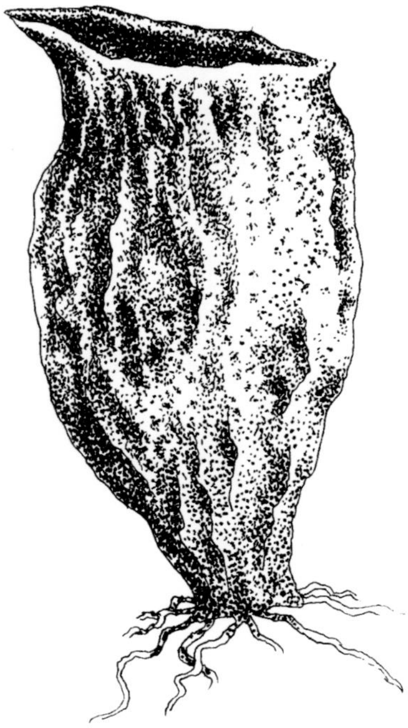

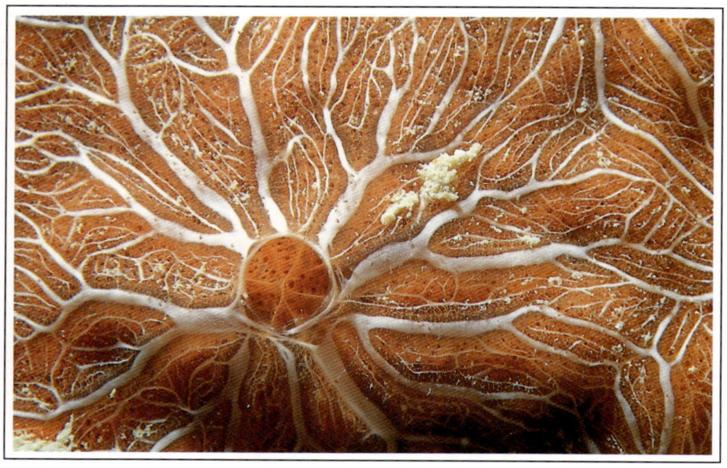

Sponges are the most primitive of all the multi-celled animals. This encrusting sponge of the class Demospongiae was growing in 15 meters off Koh Tao, Chumphon.

A tangle of tendrils, Xestospongia exigua *bears little resemblance to your common bath sponge. (Koh Krok, Pattaya, in 17 meters.)*

*Barrel sponges such as this one (*Petrosia sp.*) can grow to two meters. Barrel sponges take advantage of a chimney effect, where currents pass over the large opening, creating a suction which draws nutrient-bearing water through the millions of pores in the animal's body. (Koh Larn, Pattaya, 10 meters.)*

Many sponges grow in relatively shallow water, where wave action and currents affect their shape. One and the same species of sponge may adopt a vase-like, tabular, branching, globular or, like the specimen in this photo, a crustose form. (Koh Ba Ngu, Similans, in 10 meters.)

Many sponges are brightly colored. In some cases, the vivid hues may be a sign to potential predators that the sponge is poisonous. **Right:** *The tiny crustaceans on this specimen don't seem impressed.* **Above:** *A nudibranch (*Phyllidia sp.*) feeds on a globular sponge at a depth of 14 meters off Koh Similan.* **Below:** *This purple sponge (***below***), common in Thai waters, was growing at a depth of 9 meters off Koh Luk Ngam, Chumphon.*

COELENTERATES (phylum Cnidaria)

This very large grouping of marine organisms (about 9000 species) includes the corals, sea anemones, hydroids, and jellyfish. More complex than the sponges, they have simple digestive, muscular, and nervous organs.

Although superficially these various animals seem very different one from the other, they all share a single basic design—they are radially symmetrical and have a sac-like body with a single opening to look after feeding, elimination of wastes and, sometimes, reproduction. The opening is surrounded by tentacles bearing stinging cells (nematocysts). Two versions of this body plan have evolved: the polyp (sea anemones, corals), a cylindrical stalk bearing a mouth and tentacles, is permanently attached to the bottom; the medusa (jellyfish, for example), on the other hand, is bell-shaped and is free swimming. Some coelenterates adopt both forms—the polyp and the medusa—in the course of their life cycles. Polyps range in size from one millimeter to one meter in diameter; medusae range from being virtually microscopic to two meters.

Coelenterates have specialized tissues, but no complex organs—there are nerve cells, for example, but no concentrations of such which could be called a brain, and there is no head. In order of complexity, they seem to lie between the sponges, which consist of a variety of cells but no distinct tissues or layers, and the flatworms, which have organs made up of various tissues. It is not clear which other phyla the cnidarians might be most closely related to. There is insufficient evidence to support theories that they have evolved from colonial protozoans or else from early creatures resembling flatworms.

The classification of this phylum is a complicated one. This brief outline will look at three classes—the Hydrozoa, Scyphozoa, and Anthozoa.

a) *Hydrozoa*: This class includes the fire corals, hydroids, and siphonophores.

The fire corals (*Millepora sp.*) are often mistaken for hard coral. Appearing in a variety of forms, including encrusting and branching, this colonial animal is usually rusty colored with yellow or cream edges. If you touch it, you quickly learn why it is called fire coral. The burning and rash can last for days.

Another hydrozoan group are the sea ferns, the most common of which are white in color and grow attached to rocks and coral, waving gracefully in the currents. They also pack a sting surprising in its intensity, given how pretty and innocuous they look.

Although both the fire corals and the hydroids are commonly mistaken for true corals, they are actually more closely related to the jellyfish, and in their sexual phase they produce free-swimming medusae.

Siphonophores include the Portuguese Man-of-War (*Physalia utriculus*), a peculiar colony of specialized hydrozoans which cooperate much in the way a group of organs would in a single individual of a higher species. They can be dangerous to swimmers, with the very poisonous tentacles of the Man-of-War, for example, sometimes trailing as many as several meters behind the float. Visible on the surface is a balloon-like float with a crest which catches the wind and acts like the sail on a boat.

b) *Scyphozoa*: Jellyfish come in all sizes up to two meters across, and their tentacles may trail 30 meters behind as they drift on their mysterious migrations with the ocean currents. (They are weak swimmers, able to propel themselves by rhythmic pulsations of the bell, but relying more on currents for transportation.) They are composed of a thick layer of jelly encased in a membrane of protective cells with another, internal layer of cells making up their vital organs.

They feed on everything from plankton to large fish, and their feeding strategies vary according to species. Some move to the surface and then settle gently over their prey from above. Others feed on the bottom. One species, for example, eats while standing on its 'head', planting itself upside-down to leave its mass of feeding appendages sticking up in the current like coral, filtering the drifting plankton. Still others deploy their tentacles like spider webs or like fishermen's nets. Some conduct themselves like disks of animate flypaper, trapping plankton on their sticky surfaces.

Jellyfish are in turn preyed upon by a variety of fishes.

Contact with one of these animals can cause reactions ranging from the mildest irritation to almost instant death, depending on the species and on the susceptibility of the victim. (The most poisonous creature on Earth, in fact, is thought to be the sea wasp [order *Cubomedusae*], a variety of jellyfish found in the Indo-Pacific region, especially off the coast of northern Australia between November and April. Fortunately, they have not been reported in Thai waters.)

c) *Anthozoa*: This largest class of the cnidarians includes the hard corals, which have already been discussed in some detail. Beyond those species, it also takes in the soft corals, the sea fans, the sea pens, and the sea anemones.

The soft corals belong to the subclass of *Octocorallia*, the group of polyp animals with eight divided tentacles. (All the hard corals, the true reef-builders, are hexacorals, showing a six-sided radial symmetry, while octocorals—the soft corals, gorgonians, and black corals—are eight-sided.) Soft corals come in any number of shapes and bright colors. The octocorals do not have the reef-building limestone skeletons of their hexacoral cousins; instead they have sclerites, hard crystalline structures imbedded in their tissues. Soft, slimy, rubbery, leathery, or stiff to the touch, depending upon the species, the soft corals take a great variety of forms, including mushroom-shaped, lobed, finger-like, cauliflower-shaped, and shrublike. Because the soft coral polyps are usually visible, and because they lack the association with photosynthetic zooxanthellae, they are generally much more vivid than their hard coral cousins. Also, because the octocorals do not depend on symbiotic relationships with photosynthetic algae, they thrive at depths which do not permit hard coral growth.

Gorgonians are just one other representative of the class of *Anthozoa* which are commonly encountered in Thai waters. These corals can occur in shallow water, but tend to prefer depths of 10 meters and more. Sea whips and sea fans are two varieties, the fans in particular adding their decorative touches to the reef, growing at right angles to the current the better to collect plankton. Besides the sclerites in their tissues, they have a second skeleton of gorgonin, a hard but flexible material which gives just the right amount of rigidity to grow across the current.

Sea pens (order *Pennatulacea*) are normally only seen at night, spending the daylight hours as they do buried in the sand. They take the form of single stalks with rows of polyps arranged on them in such a way as to resemble goose-quill pens.

The second subclass of *Anthozoa* is *Zoantharia*, which includes the scleractinian corals (hard corals), the black corals, and the sea anemones.

Sea anemones are like giant flower animals, and resemble very large, single coral polyps. They can grow up to half a meter in diameter and come in a variety of colors, ranging from bright purples to pastel oranges. Though normally these creatures are sessile, many species can move slowly from place to place. Some sea anemones are known to have lived to be almost 90 years old. (However old they may actually get, they are certainly longer lived than the hydroids, some of which die and regenerate with the seasons.)

The anemones feed on a variety of fish and invertebrates. They also play host to a number of other fish and invertebrates in either commensal or mutualistic relationships—various species of anemonefish, crab, and shrimp enjoy the protection of the stinging tentacles, themselves immune because they have assimilated some of the anemone's own inhibitory chemical into their mucous coverings, tricking the anemone into accepting them as part of its own body.

Black corals (order *Antipatharia*) grow very slowly, favoring deeper waters than most corals. Their inner skeleton can be very dense, hard enough to be much in demand as a semi-precious gem material, although varieties of that quality are not found in Thai waters. These corals can take the form either of long whips or of thick bushes.

*A sea fan (*Subergorgia sp.*) found at 36 meters off Koh Miang. In depths of 25 meters of water and more, at some sites around the Similans, for example, fans grow to more than two meters across.*

*In daylight, sea pens (*Pennatula sp.*) usually remain buried in the sand. At night they emerge, revealing rows of polyps along a central stalk with the whole resembling a quill pen. (Koh Rin, Sattaheep, in 12 meters.)*

A thicket of white gorgonian sea whips and red gorgonian wire corals grow around some barrel sponges. (Koh Man Wichai, Sattaheep, 14 meters.)

*Gorgonian harp corals such as *Ctenocella sp.* do not need sunlight, and flourish at depths where they do not have to compete with hard corals. (Koh Nok, Samae San, 15 meters.)*

A specimen of Minabea sp.*, another species of soft coral, was discovered off Koh Similan in 12 meters.*

Echinogorgia sp. *are wire corals. This one, growing at 25 meters inside the wreck of the* Hardeep*, off Samae San, has its polyps extended.*

*The gorgeous animal flowers of this soft coral colony (*Scleronephthya sp.*) strain for plankton in 18 meters off Koh Rin, Sattaheep.*

Soft corals such as this Dendronephthya*—with the gorgonians, and black corals —*are Octocorallia*, eight-sided polyp animals with eight tentacles. (Koh Ba Ngu, Similans, 17 meters.)*

Above: *This* Stoichactis *is a species of carpet anemone. A couple of* Amphiprion clarkii, *or Clark's anemonefish, hover just out of the safe haven of the anemone's stinging tentacles in 9 meters off Koh Similan.* **Below, left:** *Resembling delicate flowers, tube anemones such as this one (*Cerianthus sp., *) are found on sand and mud bottoms below the reef. This long-bodied creature secretes a soft mucous tube into which it abruptly disappears, when disturbed. Cerianthids belong to the order Ceriantharia; the sea anemones proper are part of the order Actinaria. (Koh Sak, Pattaya, in 9 meters.)* **Below, right:** *A sea anemone (*Studeriotes sp.*) rests on a sandy bottom in 12 meters off Koh Miang, Similans. Sea anemones are solitary coelenterate polyps, colored in shades of orange, purple and brown, which may grow to half a meter in diameter. Generally attaching themselves to rock, coral, or mud bottoms, they are able to move slowly from place to place, although for the most part they are sessile.*

*Pink anemonefish (*Amphiprion perideraion*) such as these can live with impunity among the tentacles of the sea anemone. (Koh Ngam Yai, Chumphon, in 12 meters.)*

Radianthus sp. *are among the more common sea anemones in Thai waters. They are often found in association with* Amphiprion ephippium, *the saddleback anemonefish. (Koh Hu Yong, Similans, in 9 meters.)*

*The tentacles of the anemone appear gigantic in relation to this juvenile black saddleback anemonefish (*Amphiprion, *possibly A.* polymnus, *off Koh Rin, Sattaheep, in 18 meters).*

*These false clown anemonefish (*Amphiprion ocellaris*) and three-spot damselfish (*Dascyllus trimaculatus*) almost seem as though they have been choreographed, as they swim around sea anemones at 15 meters off Island No.5, Similans.*

Above: *Coralliomorphs (*Amplexidiscus sp.*), an order of the coelenterat sub-class Zoantharia, are represented by only a few species. Consisting o a thin basal disc covered with clubbed tentacles, they are closely related t corals, though they lack a skeleton. As do many hermatypic corals, the contain symbiotic zooxanthellae. Coralliomorphs could easily be mistaken for anemones with short tentacles. (Hin Pae, Chumphon, in 15 meters.*
Left: *Other hexacorals, the zoanthids, often flourish on intertidal zone on the reef flat, shallow areas which at low tides may be exposed to the air They are also found deeper; this one was photographed at 12 meters in Campbell's Bay, Koh Similan.*

Opposite page: *Although daisy corals belong to the hexacorals (thi specimen is probably* Tubastrea coccinea*), when their polyps are extende they might be taken instead for a soft coral. As soon as the animals retrac into their cells, however, the stony skeleton of the colony becomes apparent (Koh Ba Ngu, Similans, at night in 12 meters.)*

This Phyllorhiza punctata *was drifting near surface off Koh Chuang, Samae San. It is unwise to touch a jellyfish if you know nothing about the species at hand. It is also worth remembering, when you are in the water, that the tentacles of a jellyfish can trail far behind and below them. After storms, you can even be stung by torn-off pieces of tentacle floating free.*

Cyanea sp. *packs a painful sting. The intensity of a jellyfish sting depends on the species involved and on the extent of the contact, ranging from a scarcely noticeable sting to the potentially lethal agony of a brush with a sea wasp, or 'box jellyfish'. And the victim's recovery may take anything from a few hours to several weeks or longer. Though they are found in the Indo-Pacific region, especially off the coast of northern Australia between November and April, there have been no reports of mishaps involving sea wasps in Thai waters. (Koh Ngam Yai, Chumphon, in 12 meters.)*

*This comb jelly (*phylum Ctenophora*) ranges in size from less than 1 to 25 cm. It is not a jellyfish, being structurally more complex than the coelenterates and lacking nematocysts, or stinging cells. (Koh Rin, Sattaheep, near surface.)*

Opposite page: Palythoa memmillosa *typically grows on rocks and dead coral in shallow water. Zoanthids such as this one are related to sea anemones, which they closely resemble. The pinkish structure in the center is the gemmule, or reproductive organ, of a sponge. (Koh Ngam Yai, Chumphon, 15 meters.)*

Above: *Fire coral (*Millepora tenella*) can offer protection to a variety of small organisms, both fish and invertebrates. (Koh Similan in 15 meters.* **Left:** *This sea fern (*Lytocarpus philippinus*) is less common than its close relative* Aglaeophenie cupressina, *but it is just as unpleasant to brush against, if you aren't wearing a wetsuit. (At 21 meters on the wreck of the* Hardeep, *off Samae San.* **Below:** *Fire coral is not a coral, properly speaking; it is a hydroid. Anyone who has brushed against this innocuous looking coelenterate knows why it got the first part of its popular name, at least. It resembles hard coral, but may be distinguished in part by its mainly rust color edged with yellow or cream. (Koh Similan in 14 meters.)*

Above: *This is a detail of* Lytocarpus philippinus, *facing picture, opposite page.* **Right:** *This sea fern, a species of hydroid (*Aglaeophenie cupressina*), resembles an off-white terrestrial fern. Delicate fronds swaying in the current,* Aglaeophenie *appears altogether innocuous; but this animal packs a severe sting, and the ensuing rash may persist for some time. (Koh Ba Ngu, Similans, in 12 meters.* **Below:** *Another hydroid of the family Plumulariidae, the sea ferns, this specimen was found at 12 meters off Koh Rin, Sattaheep.*

MARINE WORMS

A piece of coral weighing just a few kilograms may harbor hundreds and hundreds of individual worms and scores of species (one study conducted in Australia found 1441 worms of 103 species in a piece of dead coral weighing something more than three kilograms).

There are both parasitic worms and independent. There are segmented worms (phylum *Annelida*), and there are unsegmented (the latter belonging to various phyla). There are flatworms (phylum *Platyhelminthes*), roundworms (phylum *Nematoda*), and ribbon worms (phylum *Nemertina*).

Some species are considerably more glamorous than your common garden variety worm, having evolved a wide variety of shapes, colors, and adaptations to different marine environments. Delicate fan worms (*Sabellid*) resemble tiny feather dusters. Or consider *Spirobranchus giganteus*—the Christmas tree worms. It would never occur to the uninitiated observer that these were worms at all. Swimming along over massive corals, you are frequently greeted with the sight of these feathery little 'trees', each 55-60mm individual a single vivid color ranging from yellow to blue to red to purple to brown to white. If you reach toward them, they abruptly disappear. Faster than the eye can follow, their feeding apparatus is sucked down into the tubes where they live embedded in the coral matrix.

Others (*Reteterebella queenslandica*) reveal themselves by extending thin white strands out of their tubes, deep in the coral matrix, for up to a meter distant, conveying little packets of sediment along these spaghettilike extrusions all the way back to their burrows. Still others (*Donellia viridis*) stretch single green tentacles for several centimeters out over the coral. As with Christmas tree worms, these various clues to the worms' existence are quickly retracted when they are disturbed.

Most worms you never see, however. Some species may be found swimming free or crawling on surfaces, but many more live tubes or else bore into sand, mud, or coral. Borers either use acid to dissolve the coral, or else grind it up with their teeth. Other species simply inhabit burrows left by other organisms.

Most marine flatworms are free-swimming members of the class *Turbellaria* (other classes of the phylum *Platyhelminthes* are usually parasitic, at least for part of their life cycles). They range in size from microscopic to more than 30 cm in length. It is thought they evolved from primitive coelenterates and, in common with them, they have a single body opening for both eating and the elimination of wastes. Eyes on found on their tentacles or along the anterior edges of the body.

Ribbon worms are more complex than flatworms. They have two openings, one at either end of the alimentary tract which runs the length of the body (30 cm or more). There is a closed circulatory system and some species have eyespots. More interestingly, they are the only animals known to have a proboscis which can be turned inside-out and which can be extended a great distance to catch prey.

Fossil burrows have been interpreted by many as evidence of wormlike creatures going back at least 620 million years. The annelid worms themselves probably evolved in the seas from an early flatworm. Most marine annelids are polychaetes—having appendages on each segment—and often have well developed nerves and muscles the lengths of their bodies, which explains the disappearing acts they can pull when approached too closely. They boast a blood system, a definite head with eyes, and a digestive tract running the length of the body.

Bristleworms, or 'fire worms' (*Amphinomidae*) are many-legged, stinging annelids found under rocks, corals, and shells, as well in crevices on the reef and in amongst algae in the shallows near the intertidal zone.

Worms play diverse roles on the reef. They provide nourishment for other organisms, for one thing. Breeding as rapidly as they do, in fact, they make a significant contribution to the general food supply. They also help break down dead coral, and thereby do their part for the continuing consolidation of the reef itself. There are even marine earthworms which closely resemble the common garden variety, and which furthermore perform the same function of aerating bottom sediments. Finally, the burrows of worms often provide commensal shelter for organisms such as crabs and gobies.

Left: *The feathery fan (4-8 cm in diameter) of this fan worm (*Sabellastrate sp.*) protrudes from a 1-cm diameter tube which extends as much as 20 cm into the sand or coral crevice. (Koh Bpida, Phuket, in 18 meters.)* **Right:** *The fan worms are segmented, and thus are members of the Annelida phylum. (Koh Chuang, Samae San, in 12 meters.)* **Below:** *Christmas tree worms (*Spirobranchus giganteus*) are also annelids. They live in tubes ranging from 1 to 8 cm in length imbedded in massive coral heads. Like the fan worm, they disappear abruptly into their tubes if disturbed. (Koh Rin, Sattaheep, in 9 meters.)*

Top: *Most flatworms live in sand or mud under stones and shells. Some—for example* Thysanozoan flavamaculatus *with its graceful undulating movement—are able to swim free. (Koh Nok, Samae San, in 15 meters.)* ***Second from top:*** *This* Baseodiscus quinquelineatus *was found under an overhang on a coral head off Samae San in about five meters. Ribbon worms belong to the same phylum as the flatworms, the Platyhelminthes, although they are anatomically rather more complex. They grow to 30 cm or more in length. Their eversible proboscis, which is employed in capturing food, is unique to the ribbon worms. (Koh Chaan, Samae San, 5 meters.* **Below left:** Thysanozoan sp. *and other flatworms, ranging from microscopic to 30 cm in length, belong to the phylum Platyhelminthes. (Koh Sak, Pattaya, in 12 meters.* **Below:** *This Pseudocerous sp. was spotted in 15 meters off Koh Larn Kai, Chumphon. Flatworms share this feature with coelenterates: a single body opening functions both to ingest food and excrete wastes.* **Opposite page:** *As unlikely as it may appear, this 'sea frost' is an annelid worm, related to the Christmas tree and fan worms. (Koh Payu, Similans, in 30 meters.)*

MOLLUSKS (Phylum *Mollusca*)

More than 100,000 different species of mollusk inhabit the reefs and coasts of the world; representative species may be found in habitats ranging from the mangrove swamps to the flats, crests, and outer slopes of the coral reefs.

This sub-kingdom includes a variety of soft-bodied, unsegmented invertebrates which usually, but not always, come equipped with hard external skeletons to which the soft body is attached by muscle tissue. The mantle, a flap of skin which is located along the edge of the shell, sets down layer after layer of calcium carbonate, extending the shell as the animal grows. Most mollusks have a ribbon-like tongue of chitinous teeth which is an effective rasping device for feeding on algae. In the cone shells, this has been modified as a tiny harpoon which is fired to impale and then inject with poison animal prey such as fish. (There are cases where people have had fatal encounters with these often beautiful but deadly shellfish.)

Overall, the mollusks come in a wide variety of body plans.

a) The largest class of mollusks are the gastropods (the 'belly-footed'), which include the snails and sea slugs. Snails feed on bacteria and algae. They may also eat coelenterates, worms, sponges, other mollusks, and even fishes. At least one species likes to dine on long-spined sea urchins. Frequently snails are scavengers, feeding on any dead organic matter they come across. Some feed upon the mucus secreted by corals when they clean themselves. Other snails—the cowries, for example—sometimes feed on the corals themselves. Movement is by wave action on the bottom of a broad fleshy foot. There is a well developed head, and a one-piece shell which is sometimes absent in the adult. Evolution appears to have tended towards the shell-less species.

Nudibranchs, or 'sea slugs', are shell-less snails. They display a great variety of form and color, with the colors sometimes warning potential predators that they are poisonous or at least bad-tasting. One species (*Phyllodesmium longicirra*) feeds on a type of soft coral which harbors zooxanthellae. The nudibranch ingests and 'farms' the algae, utilizing them as a source of energy—as organic solar batteries, in essence. Certain other nudibranchs feed on a variety of coelenterates, including stinging corals, redeploying the nematocysts and using them for their own protection. Still others eat colorful corals, distributing the pigments from the coral throughout their skin and thus adopting perfect protective coloration to match whichever host they happen to be feeding on.

b) The second largest class are the bivalves—clams, oysters, and scallops. These organisms can contribute to coral reef 'destruction', where some mussels (*Lithophaga)*, for example, bore into the hard corals. As we have seen in an earlier section, however, this sort of activity can ultimately contribute to the growth and overall health of a reef.

Bivalves feed largely on bacteria, copepods, and algae. *Tridacna*, the giant clam, also relies in part on a symbiotic relationship with *zooxanthellae*, the same photosynthetic algae which live in association with many of the hard corals.

Bivalves have a hinged, two-piece shell and most burrow in sand or mud using a strong bladelike foot. Others attach themselves to surfaces by threads of protein secreted at the base of the foot. Some (scallops [*Pecten*] and file shells [*Lima*] can swim, after a fashion, by clapping their shells to expel a jet of water. Bivalves have only a rudimentary head.

c) The cephalopods are exceptional in many ways. 'Cephalopod' means 'head-foot', and refers to the fact that the head is surrounded by a foot both divided into a circle of tentacles and modified as a funnel-shaped water-jet. The chambered nautilus has about 90 tentacles, the squid and cuttlefish 10, and the octopus eight.

They are closely related to the snails, appearances to the contrary. Hundreds of millions of years ago the ancestors of the modern-day cephalopods also had hard external shells—the chambered nautilus (N*autilus pompilius*) is one of the few surviving shelled members of this class.

Five hundred million years ago, before the emergence of the fishes, the shelled ancestors of today's cephalopods were the most successful marine carnivores. And, still, using their jet propulsion system, the squid and cuttlefish are as efficient at hunting as are most fish. (The squid and cuttlefish hunt fish, while octopuses usually feed on crustaceans.)

These creatures are in part best known for this, their chief means of getting around—jet propulsion is shared with only a few other animals, including certain bivalves. But some cephalopods, especially octopuses, are also noted for their intelligence, though their brains, the largest among the invertebrates, have evolved quite separately from those of the mammals. This may be related to the fact that, unlike most other mollusks, which can only sense light and shadow, they have excellent eyesight.

They are patient and skilful hunters, stalking prey with much evidence of deviousness, sometimes using the tips of their tentacles as lures for small fish. There have even been stories of octopuses jamming open the shells of clams using pebbles. They have been observed in the laboratory to lift the lid on their own aquarium tank, crawl across to a neighbor's, lift the cover on the second tank, eat the neighbor, and then return to their own digs—a rather macabre vignette, one worthy of a Gahan Wilson cartoon.

Highly developed camouflage and communication skills are part of the cephalopod repertoire. They can control their coloration by means of nerve cells connected from the brain to chromatophores, skin cells full of pigment. Different colors and patterns seem to be related to different conditions, for example attack, defense, alarm, camouflage, and rest states. They can even modify the texture of their skin, the better to resemble the coral they may be hiding on.

If camouflage fails when the octopus feels threatened, it can resort to ejecting a cloud of ink, an effective smokescreen to cover escape.

The squid are even further exceptional amongst the mollusks in having an efficient blood system and gills, features which are vital to the muscles and brains of these large and active animals. (Most mollusks get oxygen through a gill which is fed with blood pumped by a simple heart. Only a few species have blood colored red by haemoglobin: in the others, the blood is colorless, and relies on haemocyanin, which is less efficient a carrier of oxygen. Still, being generally smaller and less active than mammals, mollusks do quite well despite their naturally anaemic condition.) Squid and cuttlefish, like the octopus, have elaborate control over their own coloration. And the squid also uses ink jets, though in this case it appears the elongate, roughly squid-shaped ink blob hangs in the water a few moments as a diversionary tactic. While the attacker occupies itself with the dummy squid, the real McCoy turns almost transparent and makes a hasty getaway.

The cephalopods also display complex ritualized courtship and mating behavior. Some species, furthermore, take care of their young in a manner more characteristic, perhaps, of mammals.

It is not certain how *Mollusca* are related to other phyla. Partly on the basis of the undulatory motion of the basic molluscan foot, some biologists have suggested a distant connection with *Turbellaria*, a class of flatworm.

Nudibranchs are mollusks. More precisely, they are gastropods without shells—naked snails, resorting to the vernacular. They display a great variety of form and color, with the colors of species such as this Phyllidia bourguini *warning potential predators that the nudibranch is poisonous or at least bad-tasting. (Koh Ngam Yai, Chumphon, 12 meters.)*

Aeolid nudibranchs—this Flagellina rubrolineata *for example—tend to be longer and narrower in body plan than are the dorids, the other main nudibranch subdivision. Instead of forming a dorsal plume, aeolid gills take the form of cerata, or growths from the sides and back of the body which have respiratory, digestive, and defensive functions.* F. rubrolineata *does not digest the stinging cells of the hydroids upon which it feeds; rather it redeploys them throughout its cerata for defense against its own predators. (At 25 meters on the wreck of the* Hardeep, *off Samae San.)*

Casella atromarginata *is a rather elegant species of dorid nudibranch. (At 20 meters on the wreck of the* Bremen, *off Samae San.)*

Dorid nudibranchs such as these (Glosodoris sp.) *have feathery gills arrayed in a circle around the anus, on the back and towards the rear of the animal. This couple is mating. (Koh Klung Badan, Pattaya, in 11 meters.)*

Octopuses are free-swimming mollusks of the class Cephalopoda. 'Cephalopod' refers to the fact that the head is surrounded by a foot both divided into a circle of tentacles and modified as a funnel-shaped water-jet. Octopuses are among the most intelligent of animals, with a well-developed nervous system quite unlike that of the vertebrates. They can use both ink clouds and color changes as protection from predators. As shy of humans as they are, they can prove a frustrating subject for the underwater photographer. (Koh Miang, Richelieu Rock, Surins, 15 meters.)

*Squids (*Loligo sp.*) are also members of the class Cephalopoda. Whereas the octopuses have eight tentacles, both the squid and their cousins the cuttlefish have ten. The largest known invertebrate in the world is the giant squid, which can grow to more than 10 meters in length. Like the octopuses, squids secrete ink clouds, possibly as a diversionary tactic, to escape predators. Cuttlefish rely more on color change and camouflage. (Koh Similan, at night, 9 meters.)*

*The cuttlefish is another cephalopod. Cephalopods are closely related to the snails, appearances to the contrary. Hundreds of millions of years ago the ancestors of the modern-day cephalopods also had hard external shells—the chambered nautilus (*Nautilus pompilius*) is one of the few surviving shelled members of this class. (Richelieu Rock, Surins, 12 meters.)*

Top: *The spindle cowry feeds on soft corals. It often prefers varieties which are the same color as itself, its meal thus affording camouflage at the same time it provides sustenance. (Koh Ba Ngu, Similans, at night in 18 meters.)*

Middle: *Unlike many other gastropods,* Cypraea tigris *is frequently found out and about in full daylight. This habit may be contributing to the local demise of the species. Shells of the tiger cowry, many of them fist-sized, are the most commonly seen variety of cowry in souvenir and curio shops around Thailand. Understood to have special powers as amulets, some come set in thong baskets. (Hin Pae, Chumphon, in 14 meters.)*

Bottom: *Rather than grazing on algae in the manner of many other bivalves,* Tridacna squamosa *has established a symbiotic relationship with zooxanthellae. The clam harbors the photosynthetic algae within the blood sinuses of its often vividly colored mantle, which is exposed as much as possible to sunlight. These are the lurid great lips which leer at the diver from the surface of massive corals, abruptly disappearing into the stony matrix when one approaches too near. (Koh Ba Ngu, Similans, in 21 meters.)*

Top: *The giant murex (*Chicoreus ramosus*), before feeding on the soft tissues inside, uses its modified radula to bore holes through the shells of other mollusks and barnacles. In their natural habitat, murex shells are usually covered with algae. Cleaned of their dull coating, however, they can be spectacular. From the point of view of conservation, this is unfortunate, since it makes them a prize for human collectors. Hermit crabs also often favor the murex, but these crustaceans appropriate only dead specimens for use as mobile homes. (Koh Ngam Noi, Chumphon, in 15 meters.)*

Middle: *Beautifully colored and patterned as they are, the cone shells (*Conus sp.*) all share one feature which makes them highly dangerous. They can shoot out their radula, a tongue-like tooth, and inject a potent venom which paralyses their prey. Some species are lethal to humans, and one is ill-advised to handle live specimens. (Richelieu Rock, Surins, at night in 18 meters.)*

Bottom: *Oysters are mollusks of the class Bivalvia, which includes the clams, mussels, and scallops. Thorny oysters (*Spondylus sp.*) are often found in crevices or under corals on the reef slope. (Koh Similan, at night, in 15 meters.)*

CRUSTACEANS (Phylum *Arthropoda*, class *Crustacea*)

Reef crustaceans boast an incredible number and diversity of species (*Arthropoda* itself is the largest of all living phyla)—including the barnacles, shrimps, crabs, and others from microscopic copepods (tiny marine insects) to huge spiny lobsters.

Crustaceans are armored—they have a hard exoskeleton which periodically is shed to make way for new growth and the secretion of a larger shell. Lobsters, shrimps, and crabs are of the order *Decapoda*, having five pairs of jointed legs. Most in this group have two pairs of branched limbs, two pairs of antennae, and gills; the legs may be modified for swimming, walking, or feeding. Some are adapted to make sounds. The spiny lobster, for example, has at the base of each antenna a stridulator—a noise-maker which is triggered when the animal is alarmed. This may help protect it from predator fish. The pistol shrimp, on the other hand, has one large claw—its 'pistol'—with which it makes a loud snapping either to distract prey or else to startle predators.

The carapace consists of the fused head and thorax elements found in many other arthropods. The abdomen is separate and used for swimming by the lobsters and shrimps, while in the crab it has become a vestigial flap on the underside. Crustaceans have compound eyes. Reproduction is sexual, with the male depositing sperm on or near the opening of the female's oviduct. Most decapods carry the embryos till they hatch.

Crustaceans are omnivorous, eating plants, animals, and sometimes filter feeding on organic matter as well.

Often fantastically formed, vividly colored and patterned, they are masters of camouflage, either using bits of the environment—sponge, coral, sand, and other rubble—or else adopting colorations which blend in with coral or urchin backgrounds. During the day, however, these animals are not always in evidence, no matter how hard you look, since a good number of species hide from predators in daylight and only emerge after nightfall to forage for food.

Unlike most of their temperate and cold-water cousins, many coral reef crustaceans enter into symbiotic relationships with other organisms. It is for this reason—the mutualistic relationships with a variety of coelenterates, mollusks, and echinoderms—that there is such a bewildering variety of crustaceans on the reef; the various species each have their own little niche within the overall ecosystem, and thus do not exert undue competitive pressures on one another.

Possum shrimps (*Heteromysis harpaxoides*), for example, live inside gastropod shells already inhabited by hermit crabs, often in families of a male, a female, and young. They seem to feed on waste matter, keeping the squat clean for one and all. Pea crabs (*Xanthasia murigera*), on the other hand, live in cozy domestic pairs inside the mantle cavity of the giant clam, one pair to a clam. Intruder crabs are probably eaten. One mutualism very commonly observed in the Indo-Pacific involves a little fish called a goby and a shrimp which share a burrow in the sand bottom. The goby stands guard at the entrance; if you watch carefully, soon you will see the shrimp appear, clearing sand and pebbles away with its pincers. Disturbed, the flatmates both abruptly make themselves scarce. Only when the goby has once more stood watch will the poorly sighted shrimp resume its excavations. There are many other examples possible, including the arrangement the pistol shrimp and xanthid crabs have with the shrub coral, exchanging protection from fish predators for protection from crown-of-thorns attacks (as we saw in Part One, above). The cleaner shrimp have also been discussed already (Part One, above).

Other shrimp, for example *Gnathophyllum americanum*, prey on starfish, including the crown-of-thorns, and may help in keeping reef predation down.

This crustacean has been only tentatively identified as a spider crab, miniature crustaceans often found on sea fans and soft corals at night. (Koh Ba Ngu, Similans, in 17 meters.)

This 'decorator crab' (Camposcia sp.), discovered in 15 meters on the wreck of the Hardeep *at night, is camouflaged with algae and bits of shell.*

This crustacean is probably a species of Majidae. (Koh Sak, Pattaya, in 15 meters.)

*The painted reef lobster (*Panulirus longipes*) keeps itself scarce during the day, itself not keen on the idea of Lobster Thermidor. (Koh Miang, Similans, at night in 12 meters.)*

Naturally camouflaged, this crab is all but indistinguishable from the sponges, tunicates, and algae encrusting the reef substrate. (Koh Ba Ngu, Similans, at night in 12 meters.)

*Swimming crabs (*Thalamita sp.*) sometimes hitch rides on jellyfish. More often they may be seen swimming near surface. (Koh Nok, Samae San, in 15 meters.)*

Periclimenes sp. *is commonly called a 'ghost shrimp, for obvious reasons. (Koh Ba Ngu, Similans, in 20 meters.)*

BRYOZOANS (Phylum *Bryozoa*)

Bryozoa is thought to include several hundred species, though research into these reef animals has not been very extensive. Popular names for these creatures include 'moss animals' and 'sea mats'. Bryozoans are usually sessile, tending to seek shaded areas, growing on rocks, shells, crustaceans, and any other handy surfaces. Studies have shown they are often the first animals to move into the reef environment, even enjoying a period of relative dominance.

It was once believed that bryozoans were seaweeds. Later they were mistakenly classified as hydroids since, like hydroids, most bryozoans are colonial. The colonies adopt a variety of forms, including small, fern-like dead ringers for certain hydroids. Others establish themselves as red calcareous crusts, while the most striking variety is sometimes misleadingly called 'lace coral', and appears on reef ledges as a fine white lace. Some colonies form spherical masses of up to 200 mm in diameter, each new layer of individuals growing on top of and killing the one underneath, the colony as a whole gradually expanding like a balloon.

The individual animals are very small, and the entire colony is frequently overlooked by divers. Viewed under a microscope, however, they can be quite beautiful.

The bryozoans have been around for at least 500 million years. Each individual is little more than a digestive tube with mouth and anus, the rigid outer wall consisting either of chitinous tissue or incorporating calcium carbonate. Filter feeding is effected by means of a ring of ciliated tentacles around the mouth.

The bryozoans themselves are fed upon by some species of dorid nudibranch.

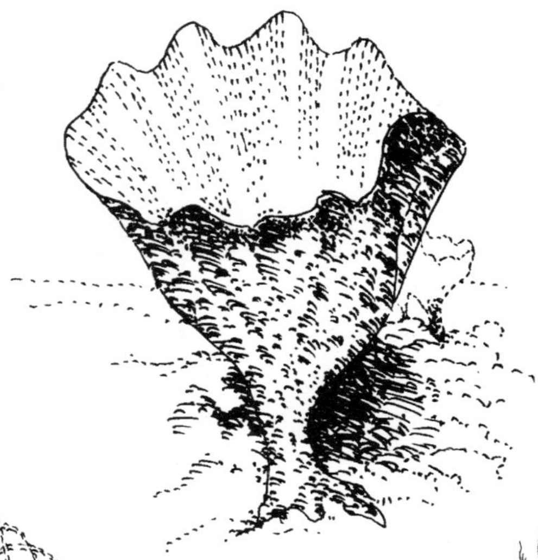

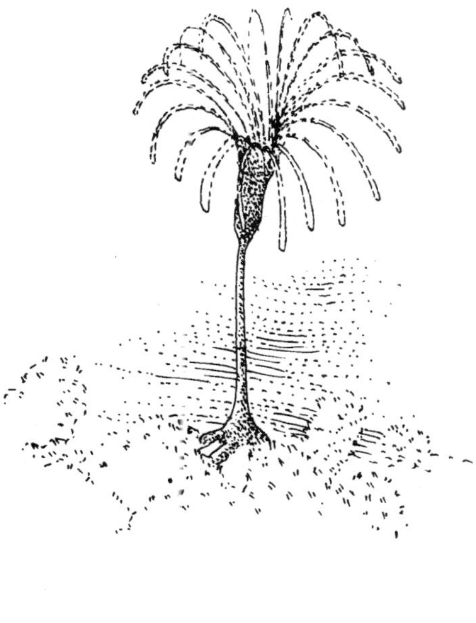

*This bryozoan (*Reteporellina sp.*) takes a lacy form. Other species appear as red calcareous encrustations on shellfish or as fine, fern-like growths which could be mistaken for tiny hydroids. Historically, these colonial animals were mistaken first for plants and then for hydroids. (Koh Similan in 12 meters.)*

ECHINODERMS (Phylum *Echinodermata*)

Five classes of echinoderms are found associated with coral reefs:

1. sea stars (*Asteroidea*)
2. sea urchins (*Echinoidea*)
3. brittle stars (*Ophiuroidea*)
4. sea cucumbers (*Holothurioidea*)
5. feather stars (*Crinoidea*)

Most echinoderms are bottom dwellers in shallow waters. All echinoderms display a five-part radial symmetry around a central mouth (overlaying a more basic bilateral symmetry), and they have an internal skeleton made up of many calcium carbonate plates which is usually spiny and which knobbles the surface of the animal, giving the animal a thorny or warty appearance (the name 'echinoderm' is made up of words meaning 'hedgehog' and 'skin'). The body cavity is a water-vascular system, an internal hydraulic system where fluid-filled vessels are extended and used in locomotion, feeding, respiration, and sensory perception (being sensitive to touch as well as changing intensities of light and temperature).

These animals contribute to the reef community partly as scavengers of decaying matter and partly by regulating numbers in other species upon which they prey. Their larvae, great numbers of them, go towards feeding planktonic organisms.

Holothurin, a toxin derived from some sea cucumbers and used in some places to kill fish, has been shown to inhibit the growth of some tumors in mammals.

Most sea stars have a relatively smaller central part, with a mouth on the undersurface and an anus on the upper. The long, slender arms radiate from the body. In others, such as the box stars, the arms are so merged with the body that they simply form the tips of a boxy pentagon. The arms have tubular feet which allow the animal to adhere to surfaces and resist the forces of wave action. These tube feet may also help in finding food. When feeding on bivalves, the starfish attaches itself to the valves with its tube feet and then raises the center of its body, exerting a remorseless upward pressure which gradually prises the shell open. Finally, the starfish's stomach is extruded to digest the shellfish *in situ*.

The crown-of-thorns (*Acanthaster planci*) is the most well-known, or perhaps most infamous, contributor among the echinoderms to the evolution of the coral reef. The crown-of-thorns enjoys a worldwide notoriety, often being cast as the villain when there is talk of reef destruction. Nevertheless, they have a certain strange beauty, and it has by no means been resolved whether or not they are in fact part of natural cycles which ensure the health of the reef over time.

Generally feeding at night, the crown-of-thorns distends its stomach over a patch of coral to secrete digestive juices which turn the coral polyps into a convenient and no doubt tasty soup. In one night, a single starfish can digest an area about half its own size—this amounts to five square meters or more every year.

These creatures tend to congregate in groups, attracted by the feeding juices of their own kind. One female can produce 20 million eggs per year. Where there is a real population boom, and you get thousands of individuals on a reef, up to 90 percent of the coral may be killed over a given area.

The sea urchins are the largest single source of continuing erosion through grazing on hard corals. Long-spined Black Sea Urchins (*Diadema setosum*) are the most commonly seen in Thai waters, often congregating on sand surfaces surrounded by corals and rocks. Like the sea stars, most sea urchins have a mouth on the undersurface and an anus on the upper. Others, such as the Burrowing Heart and the Sand Urchin, for example, have both on the underside. Polychaete worms, some fish, and crustaceans are frequently found in commensal relationships with urchins.

The brittle stars have a single opening on the underside which serves both as mouth and as anus. The digestive and reproductive organs are located in the relatively small central disk, and do not extend into the arms, as they do in the starfish. Growing to about 30 cm in diameter, these spidery echinoderms are most active at night.

At first glance, sea cucumbers do not seem to have a lot in common with starfish or sea urchins. The five-rayed symmetry is still there, however, despite the rather worm-like body: five bands of tube feet run the length of the body. The typical echinoderm skeleton exists only in vestigial form as microscopic spicules imbedded in the body tissues. Shallow-water dwellers, ranging in size from a couple of centimeters to a meter and more, they extend tentacles to sweep the bottom surface or else to trap floating particles in the water. Some sea cucumbers host little fish, worms, and pea crabs which enjoy a cozy and quite secure home, complete with constantly circulating oxygen-rich water, in the holothurians' cloacae.

The feather stars are the oldest and most primitive class of echinoderm. At one stage, 300 million years ago, they were in fact the dominant animal in the sea. Although they don't generally impress one with being the masters of the deep, they do add a nicely decorative touch to the reef. They are often found in clusters at strategic vantage points such as along the ridges of spur-and-groove systems on a reef slope, or on reef walls where plankton is abundant. At night, they often climb up on the edges of fan corals and other prominent locations. Relying on the currents to bring food within range, each feather star, depending on its species, fans anywhere from five to 200 arms out across the current in a layered baffle, very effectively straining the water for plankton. The strategy must be successful—this creature has survived largely unchanged for more than 400 million years.

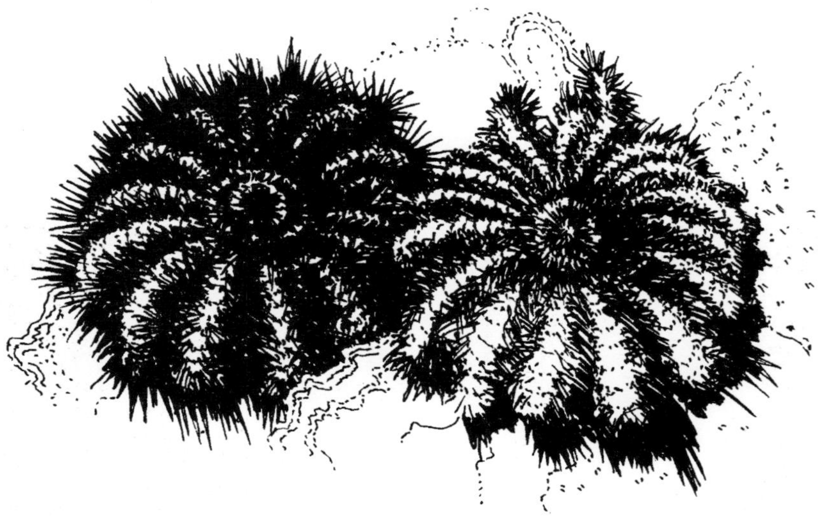

Above: *The feather stars, or crinoids, are the most primitive of the echinoderms. Like related species,* Comatella maculata, *a brilliant red ornament on the reef, has remained essentially unchanged for the past 400,000,000 years. Some 300,000,000 years ago, the crinoids were the dominant animal in the sea. (Koh Racha Noi, Phuket, in 18 meters.* **Below:** *The arms of feather stars such as* Comatella nigra *are covered with a poisonous mucus. Any plankton which is captured and incapacitated is then passed to the central mouth along grooves on the arms. (Koh Hu Yong, Similans, in 11 meters.)*

Above: *Cushion sea star arms have merged, the five tips of the pentagona body their only vestige. In this photo,* Culcita novaeguineae *is attached t a rock at 15 meters off Koh Ngarm Noi, Chumphon.* **Left:** *Aside fron using their tubular feet to move in any direction, the sea stars also use then as anchors and in preying on other organisms. (*Linckia laevigata *off Islan No.6, Similans, in 18 meters.)* **Below:** *Sea stars prefer muddy substrate and rocky or coral reefs, having adopted the same habitats as those of thei main diet, the mollusks. (*Nardoa sp. *off Koh Similan in 12-18 meters.)*

Oligometra serripinna *is another striking crinoid species. At night, feather stars move to such prominent vantage points as the edge of fan corals. There they array anywhere from five to 200 arms in a layered baffle across the current, straining for plankton. (Koh Racha Noi, Phuket, in 18 meters.)*

Above: *This variety of crown-of-thorns (*Acanthaster planci*) displays a coloration specific to the Andaman Sea, one distinct from the brownish-orange common to crown-of-thorns in the Gulf of Thailand. (Koh Miang, Similans, at night in 15 meters.)* **Opposite page:** Fromia monilis, *this one found in 9 meters off Koh Payu, Similans, is a beautifully colored and patterned sea star.*
Below: *Sea stars are the most widely recognized of the echinoderms. Their basic five-membered radial symmetry means that most sea stars have five arms radiating from a central body area, although in some species the number of arms may vary from four to forty. (*Pentaceraster regulus *Koh Nok, Samae San, in 14 meters.)*

Top: *The long-spined black sea urchin (*Diadema setosum*) is that one most commonly encountered in Thailand's coastal waters. Abundant on reef flat and slope, usually found on sand surrounded by rocks and coral,* Diadema *is a beautiful creature, resembling a strangely jewelled pin cushion, though its 'pins' may be 30 centimeters long. It is not dangerously poisonous to humans. If stabbed, the victim can adopt the local remedy and simply whack at the puncture wounds with a rock or knife handle till the broken-off tips of the spines have been crushed. The body will assimilate the spine tips, and any pain should go away soon. (Koh Sak, Pattaya, in 15 meters.)* **Middle:** *Sea urchins such as this* Mespilia globulus *are capable of moving about the reef as they feed on algae and coral. Aside from people, their chief natural enemy are the starfish, fellow echinoderms though they may be. (Richelieu Rock, near Koh Surin, in 15 meters.)* **Bottom:** *The debris littering this specimen of* Toxopneustes pileous *is not accidental in origin. Whereas, during the day, many other species of sea urchin tend to hide under coral heads and rocky overhangs,* Toxopneustes *instead uses bits of stone and coral to camouflage itself. This sea urchin is poisonous, so resist any impulse to pet it. (Koh Nok, Samae San, in 15 meters.)*

Above: *The body of a brittle star may be only a couple of centimeters in diameter, while its arms can extend 30 cm or more across. These echinoderms got their common name because their long arms break off so easily (although these can be regenerated). These specimen are clinging to a gorgonian coral off Koh Ngam Yai at 14 meters.)* **Right:** *Ophiothrix sp. is seem here hitching a ride with a jellyfish near surface off Koh Sak, Pattaya.* **Below:** *Closely related to the brittle stars are the so-called basket stars. This one was photographed off Koh Rin, Sattaheep, in 17 meters.*

A brittle star (Ophiothrix purpurea) *has attached itself to a gaudy* Dendronephthya *soft coral in 14 meters at night off Koh Similan.*

Above: *This sea cucumber (*Holothuria leucospilota*) is common in Thai waters, especially on sandy bottoms around shallow reef flats. In the photo it is spawning. When alarmed, some sea cucumbers are able to eviscerate themselves, expelling their internal organs through the cloaca. This may be a strategy designed to preoccupy predators. The organs are regenerated within several weeks. Some sea cucumbers are also able to extrude long sticky threads which can entangle attackers. (Koh Ngam Yai, Chumphon, in 15 meters.* **Bottom:** Thelenota ananas *is a spiny species of sea cucumber. This one is feeding in 27 meters off Richelieu Rock, near Koh Surin. Sea cucumbers are directly threatened by humans only to the extent that they are a prized ingredient in Chinese cuisine. They are also reputed, in parts of Asia, to have aphrodisiacal powers. Medical researchers, meanwhile, are investigating the sea cucumber toxin holothurin as a means of inhibiting the growth of tumors. (***Inset:** *detail of* T. ananas.*)*

ASCIDIANS (Phylum *Chordata*, subphylum *Urochordatga*)

These animals are more commonly known either as sea squirts or tunicates. 'Ascidian' is derived from *askos*, the Greek for 'wine flask', and very often they do resemble leathery little bags with two openings—one to let in water bearing oxygen and plankton, and one to let waste water out. They can close one or both openings, but more often leave both open, with a stream of water passing continuously through. A 30-mm sea squirt can filter about one liter per hour (they range from several millimeters to several centimeters in size).

However primitive this description may make the sea squirts seem, they do in fact have stomachs, intestines and, being hermaphroditic, not only a testicle but an ovary too. The test surrounding these animals is supplied with blood vessels. The mantle has muscles by means of which the tunicate can open and close its siphons or contract the body as a whole.

Ascidians belong to the Phylum *Chordata*. Although they do not have a backbone, their larvae have notochords which help support muscular tails and which are evolutionary harbingers of the true backbones. The larvae also have gill slits, which would also seem to ally them more closely than their adult forms to the higher organisms. It is indeed thought possible that the tunicates are the ancestral chordate—ancestor, that is, to all of us.

In their adult phase the tunicates become sedentary (much like many humans), attaching themselves either singly or in colonies to the undersurface of rocks on the reef flat, to the roofs and walls of caves on the reef slope, and to the sides of holes and crevices. They also make some contribution to the consolidation of rubble caught in depressions around the reef rim (although ascidians can be found everywhere from the low tide mark to the ocean deeps as well).

Their social life is more interesting than one might grant at first glance. Besides propagating hermaphroditically, for example, some reproduce by budding. And individuals sometimes cluster around a single common cloaca, thus constituting 'composite ascidians'. So-called 'social ascidians', on the other hand, remain united to the parent individual after budding.

Sea squirts are protected from humans by the fact they aren't much good for anything (or so it was believed until it was discovered that some yield substances, such as didemnin B, which are helpful in the treatment of leukemia and tumorous growths). Their tough, leathery bags protect them from a variety of other predators, however, notably starfish and various species of mollusk. The sessile varieties, though, are often displaced by more competitive organisms such as sponges. In at least one instance, tunicates offer protective camouflage to another animal—some types of crab characteristically carry an envelope of sea squirts which may help protect the crustacean from its enemies.

There is a class of tunicates (*Thaliacea*) which are free-swimming. Ranging in length from a few centimeters to two meters in length, these animals are transparent, cylindrical or cone-shaped creatures sometimes mistaken for jellyfish.

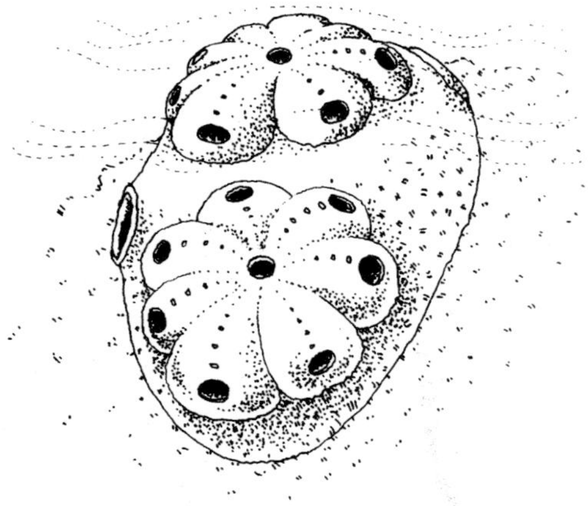

This encrusting, colonial species of tunicate (Clavelina sp.) was found in 12 meters off Koh Dok Mai, Phuket.

Tunicates, as primitive as they may appear, are of the Phylum Chordata, and may indeed be the ancestral chordate—ancestor to all of us. (Koh Similan in 12 meters.)

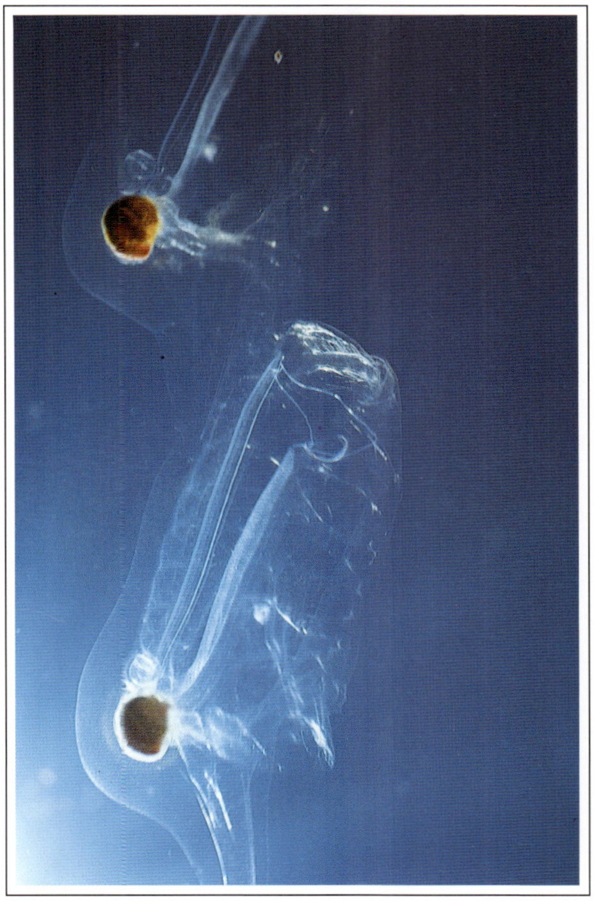

Often mistaken for a jellyfish, this is really a free-swimming tunicate (Thalia sp.). *(Koh Miang, Similans, near surface.)*

This tunicate (Polycarpa sp.) *is of a solitary species, seen here surrounded by anemones. (Koh Man Wichai, Sattaheep, in 15 meters.)*

CORAL REEF FISH (Phylum *Vertebrata*, class *Pisces*)

A greater number of fish species are found on tropical reefs than anywhere else in the seas. On the Great Barrier Reef, for example, 150 species have been counted in a single 100^2 meter area. In Thai waters, it is estimated that well over 100 species of fish make their homes on the coral reef.

Today there are two main classes of fish: those with cartilaginous skeletons, including sharks and rays, and those with bony skeletons, which represent most others. They range in size from gobies which, at only 20mm and weighing a gram or so, are thought to be the smallest living vertebrates in the world, to groupers weighing hundreds of kilos to the magnificent whale shark (*Rhincodon typus*) which reaches lengths of 18 meters and weights of more than 15 tons (though the latter is not really a reef fish, it does visit these coastal waters occasionally).

These cold-blooded animals have evolved to fit almost every aquatic niche there is. Beyond that, it has been pointed out that all land vertebrates may fairly be described as highly modified fishes.

Though their evolutionary origins are shrouded in mystery, the jawless fishes of some 450 million years ago probably arose from soft-bodied creatures similar to the bottom-dwelling, filter-feeding *Cephalochordata* (*Amphioxus* and related organisms), small eel-like animals with no true head or backbone, but with features which indicate they are nevertheless the forerunners of vertebrates.

The most striking thing about the reef fish, in the popular mind, is their brilliant color and pattern and variety. In fact, there are few things on Earth which can so dazzle the senses. But *why* are they so vividly turned out?

In the first place, the colors and patterns help individuals recognize others of the same species—from an adaptive point of view, it would be wasteful to try reproducing with the wrong species. And species recognition is important for another reason, according to some scientists, inasmuch as the reef is divided up into territories claimed by various individuals, and these gentrified fish need to fight off trespassers of the same species. (Fish of other species can often be tolerated, on the other hand, since they are not always in direct competition for the ecological niche—many and subtle are the differences in chosen habitat and preferred food among the species.) In other cases, in their natural habitat and against the background of a riot of similarly striking colors and patterns, the vivid hues of the reef fish often actually provide camouflage. Finally, where the fish does in fact stand out, evolution may have provided it with other survival tactics. Some butterflyfish, for example, have dark stripes which conceal the real eye, and very distinct but false eye spots at their other end; this can trick a predator into attacking the wrong end of the body.

The shapes of reef fish have also been determined by the demands of survival in this particular environment, with different species adapting to the various niches offered on the reef. Whereas open-water fish such as tuna and marlin are designed primarily for speed, with a long, hydrodynamic shape tapering from head to tail, many reef fish are instead designed for agility. Disc-shaped species such as the surgeonfishes, butterflyfishes, and angelfishes are adept at maneuvering in the close quarters of the reef, able to back up and turn in narrow crevices and holes when hunting food or avoiding predators. These fish are laterally compressed, deep, short, and have pectoral fins high on the sides of the body. When the fins are extended, the fish can pivot around them, turning or stopping on a dime. Other coral reef fish—scorpionfishes and lizardfishes, for example—spend most of their time lying flat on sand or rock or coral, and have bodies which are broader on the bottom than on top, affording a stable base. Their fins, at the same time, have become thick and fleshy and are used like rudimentary legs. Others still—the trumpetfish and needlefish or example—have evolved long, thin shapes which on one hand allow them maximum flexibility in the crevices of the reef, and on the other hand, together with their colors, make them difficult to see for predator and prey alike.

The fishes contribute to the reef ecosystem chiefly through their places in the food chain. The algaes, sea grasses and plankton are the primary producers, but they are constantly grazed upon by fish and other organisms such as crustaceans and echinoderms which are in turn preyed upon by various fish. The fish, in their turn, contribute to the common larder.

The hard corals, for instance, grow better where fish school above them, flourishing in the constant rain of excreta; even after digestion, the foodstuff taken in by the fish carries valuable nutrients. And, after a sequence of 'big fish eats littler fish' incidents, the largest organism in the chain (a shark or a grouper, for instance) will eventually die, after which smaller creatures will have their day, getting to feed in turn on the larger predator. Bacterial action will break the remaining organic matter down into its constituent minerals, and the algae will once again take these up, incorporating the material once more into the food chain.

Finally, there is at least one more way in which the fishes contribute to the reef community—they have evolved together with the corals and with other animal species on the reef, establishing mutualisms of a bewildering number and diversity.

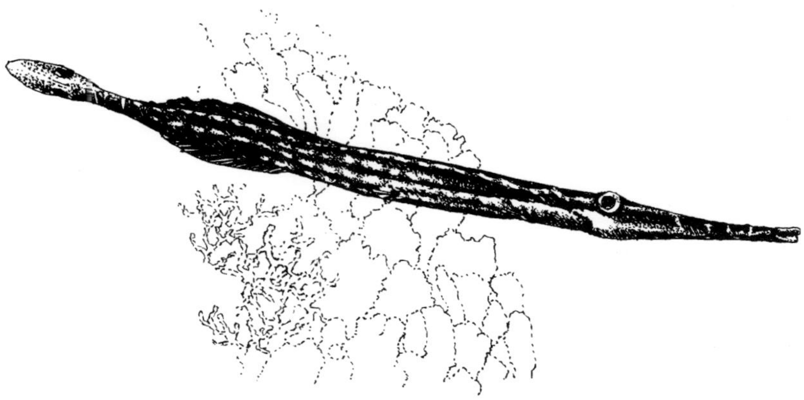

Above: *A moray eel (*Gymnothorax thyrsoides*) leers at an intruder in 14 meters off Koh Man Wichai, Sattaheep.* **Below:** *Although this blue-spotted fantail stingray (*Taeniura lymma*) appears to be rounding on the camera, and even though it is capable of delivering a very painful sting with its tail, it is really a timid creature. (Koh Rin, Sattaheep, in 12 meters.)* **Left:** *Species such as the gold-striped goatfish (*Mulloides vanicolensis*) are often seen swimming solitary or in small schools around the reef edge or along ledges on the slope. (Koh Hu Yong, Similans, in 6 meters.)*

*Although the manta ray (*Manta hamiltoni*) is not a regular inhabitant of the coral reef, it is an occasional visitor. Despite its size—a manta can run to more than six meters in breadth and more than 1000 kilos—and in spite of what is, to some, its fearsome appearance, the manta is harmless. The tail, unlike that of the stingray, is not poisonous. The distinctive forward appendages are used as an aid in feeding. At least 10 varieties of manta have been reported in the Indo-Pacific region, though they may all belong to one species. Unlike most rays, which live near the bottom, mantas hunt near the surface. (Koh Similan, Fantasy Reef, in 12 meters.)*

*Leopard sharks (*Stegostoma fasciatum*) are for scuba divers the main attraction of Shark Point, just off Phuket. These docile creatures (up to about 3 meters in length) allow people to come right up and hug them with impunity. Unfortunately, there has been too much of this sort of interference, and the sharks are beginning to move off to other places. (Koh Tachai, north of the Similans, in 30 meters.)*

*Nurse sharks (*Nebrius sp.*) are common around areas such as Pattaya, the Similans, and the Burma Banks. Less aggressive than many other species, they even permit divers to play with them. Nevertheless, take care not to annoy or to corner one—they have been known to bite onto a diver's arm and refuse to let go. This could prove embarrassing, if one were running low on air. (Koh Surin in 15 meters.)*

A goby like this Hoplichthys *(possibly* H. gilberti*), at 2 cm long, is at the other end of the scale from the whale shark. (Koh Similan, at night, in 14 meters.)*

*This scorpionfish (*Inimicus sp.*, possibly* I. didactylus*) was photographed off Koh Larn, Pattaya, in about 15 meters. Masters of camouflage that they are, it can be virtually impossible to distinguish one of these fish against a background of coral and algae—at least till an eyeball moves and the Gestalt falls into place. Scorpionfish are poor swimmers, their body shape adapted instead to lightning strikes from ambush at the small fish which make up their diet.*

Above: *The zebrafish (*Pterois volitans*) is also known as the 'firefish', and with good cause. Zebrafish are fearless, and will allow a diver to approach close enough to pet them. From the diver's point of view, however, this would be a serious mistake. Having come in contact with the poisonous dorsal spines, the victim experiences immediate and intense pain, which quickly radiates out from the wound to the rest of the body. With most varieties of zebrafish, the pain begins to disappear after a few hours. (Koh Miang, Similans, in 15 meters.)* **Middle:** *Among the most striking of reef fish, powder-blue surgeonfish (*Acanthurus leucosternon*) swim in schools of from a few individuals to a couple of dozen and more. They are found around rocks and coral in depths of anywhere from 30 meters and deeper to just a few meters from surface. 'Tangs', as they are sometimes called, scrape and bite algae from rocks and coral. When threatened, surgeonfish erect razor-sharp 'scalpels' on either side of the tail, weapons with which to slash at enemies as they flash past. (Koh Similan, Campbell's Bay, at night in 11 meters.)* **Bottom:** *The porcupine pufferfish (*Diodon liturosus*) is easy to catch at night. A favorite sport for divers is to capture one and watch it blow itself up into a spiked balloon, letting it bobble away, then, its fins fluttering frantically in the light of an underwater torch. This practice is ill advised, however, since it can seriously injure the fish. (Koh Similan, at night, in 12 meters.)*

Amblyglyphidodon aureus, *or the golden damselfish, is dwarfed against the stalks of a* Dendronephthya *soft coral. (Koh Ba Ngu, Similans, at night in 15 meters.)*

Diademichthys lineatus, *the long-snout clingfish, typically relies on the spines of sea urchins* (Diadema sp.) *for protection. (Koh Ngam Yai, Chumphon, in 15 meters.)*

Nothing shows but the head of this goby (Aspidontus sp., *possibly* A. taeniatus) *as it peers cautiously from its hole in the coral. (Hin Pousar, Similans, in 11 meters.)*

Delicately hued blue-green chromis (Chromis caerulea), *one of the many species commonly referred to as damselfish, hang in schools around shallow-water branching corals. (Koh Miang, Similans, in 11 meters.)*

*This boxfish (*Ostracion cubicus*) is a poor swimmer, and not nearly so big as it would have you believe, when you encounter it in one of its hiding places. Given the size of the head—usually all that is visible when the fish is hunkered down in a cave—and the prominence of the forward-looking eyes, you get the initial impression that you dealing with something about three meters long. Only when* Ostracion *finally makes a break for it do you see the absurdly short body which lurked behind the head. (Koh Similan, at night, in 12 meters.)*

*Like the other triggerfish, the titan or blacktipped triggerfish (*Balistoides viridescens*) frequents shallow waters, grazing on algae in seagrass beds as well as both*

*The spotted boxfish (*Ostracion meleagris meleagris*) is another familiar face around Thailand's coral reefs. (Koh Similan 12 meters.)*

*Three-spot damselfish (*Dascyllus crimaculatus*) are designed such that they can swim quickly in and out of the coral thickets from which they never stray far. (Koh Hu Yong, Similans, in 13 meters.)*

*This is the scribbled leatherjacket (*Aluterus scripta*). Enjoy the fish for its color and its whimsical design, but don't plan to enjoy it for breakfast—like many reef creatures, it is poisonous. (Koh Miang, Similans, at night in 18 meters.)*

Cephalopholis miniatus *is a small grouper common on Thai reefs. Various species go by the names 'coral trout' or 'coral rock cod'. (Koh Similan in 15 meters.)*

*Blue-spotted fantail rays (*Taeniura lymma*) are frequently encountered around shallow coral heads and rocks, where they are quick to find cover in caves or crevices. Shy as they are, you wouldn't want to have one sting you. (At 27 meters on the wreck of the* Hardeep, *Samae San.)*

*The regal angelfish (*Pygoplites diacanthus*) favors small caves and ledges on the reef. Disc-shaped, this often solitary fish is designed for quick maneuvers in close quarters. (Koh Similan in 15 meters.)*

Spotfin lionfish (Pterois antennata), *like their cousins the zebrafish, conceal poisonous spines among their feathery dorsal fins. You can come right up to a specimen such as this one, hovering at 15 meters off Koh Miang, Similans.* Pterois *shows no alarm whatsoever, beyond gently inclining so that its dorsal spines are tilted towards any intruder who is too persistent.*

Above: *The striped barracuda* (Sphyraena obtusata) *is a fast, streamlined fish which hunts in large schools near the surface. Though barracuda are basically pelagic, or open-water fish, they are often found in smaller schools over reefs. (Koh Miang, Similans, in 9 meters.)*

The variegated lizardfish (Synodus variegatus), *like other lizardfish, spends most of its time lying flat on the bottom. This solitary species has evolved a body shape which is broader on the bottom than on top, affording a stable base. Its fins, meanwhile, have become thick and fleshy, the better for use as rudimentary legs. (Koh Similan in 12 meters.)*

Butterflyfish have the deep and compressed bodies typical of other reef dwellers which rely for survival on agility in close quarters. They have a second line of defense, as well: by erecting their dorsal and anal spines, they can wedge themselves securely into place in a crevice. Some species have yet another strategy—false eyespots may prove confusing to predators, sometimes tempting them to attack the wrong end of their prey. The butterflyfish themselves feed on coral polyps, other invertebrates, and algae. Their colorful appeal has made them prime targets for aquarium fish collectors. (Meyers butterflyfish [Chaetodon meyeri]; *Koh Similan, at night, in 9 meters.)*

Trumpetfish (Aulostomus chinensis) *display yet another standard body type, this one differently adapted to life on the reef. Evolution, in animals such as the trumpetfish and needlefish, has come up with a natural 'stealth' technology. On one hand, their long, thin shape allows them maximum flexibility in the crevices of the reef. On the other hand this shape, together with their usual color, silver-grey, makes them difficult to see for predator and prey alike as they hunt near the surface. (Koh Hu Yong, Similans, in 14 meters.)*

MARINE REPTILES (Phylum *Vertebrata*, class *Reptilia*)

Reptiles emerged about 280 million years ago, somewhere between amphibians and birds, with birds and mammals both having since evolved from reptilian ancestors. Even the marine reptiles have lungs, which testify to the emergence of their earlier relatives from the seas to live on land, some time before they reverted to an aquatic existence.

Sea snakes (sub-family *Hydrophidae*)—the banded coral snake is the most common species in Thailand—have evolved a paddle-shaped tail, to aid in swimming. They also have valved nostrils which allow them to seal off breathing passages against the water when submerged. They generally surface every 20-30 minutes to breathe, though they can stay down an hour or more.

All marine snakes are poisonous: some of them are much more deadly than the king cobra. Fortunately, they are normally also shy, and leave divers alone. Even if they are alarmed to the point they bite, according to some observers, their mouths and their fangs are generally so small that they can't get a purchase on anything other than an earlobe or the web between one's fingers. Only the very largest can bite through a wetsuit.

Few species ever come ashore, and sea snakes are the only marine reptiles which actually bear their eggs or their young in the water. These snakes often live in colonies near the shore, and may be seen sometimes on the reef flat and on the reef slope. More than 30 species have been identified in Thai waters.

The hawksbill is the most commonly sighted sea turtle in Thailand. This species is largely herbivorous. Its name refers to the fact that both its shell and its meat are green. It usually runs to 70 kilos, though some individuals may weigh as much as 365 kilos. Being reptiles, they have a low metabolic rate; and it is thought these animals use 300-400 times less oxygen, in proportion to their size, than do humans. This allows them to stay submerged for long periods. Despite their size and low metabolism, furthermore, they are capable of swimming 300 feet in 10 seconds, a capacity which is obvious to every scuba-diver who tries to hitch a ride on one. (In fact, attempted hitchhiking on turtles is a poor idea, since the stress occasioned by these chases can be harmful to the turtle.)

Sea turtles are something of a mystery—little is known of their life-cycles outside the periods when they have just hatched and when the female comes ashore to lay her up-to-200 eggs. (Almost nothing is known of the male, since it never comes on land.) It has been determined that sea turtles commonly make voyages of 1000 km or more in a single year, however. It is also believed that turtles always try to return to breed on the islands upon which they were born. How they navigate on these journeys is open to question. Suggestions that they find their way by the sun and stars are countered by the fact they are very myopic in air and, according to studies done, can only see stars from beneath the surface in extremely calm conditions. A more likely factor guiding their migrations, in some opinion, is the ability to recognize and follow in the ocean currents minute quantities of trace compounds which identify their homes.

Given the abandon with which these creatures are hunted for their meat and shells, and given the continuing disruption of their breeding grounds, there is a very real danger the full story will forever remain a mystery. Until recently, there were five species of sea turtle in Thai waters: the loggerhead, the leatherback, the green, the hawksbill, and the Pacific Ridley. The loggerhead is already locally extinct, while the green is considered rare.

The hawksbill turtle (Eretmochelys imbricata), which is becoming rare in Thai waters, is now on the endangered species list. Not only is it hunted for shell and meat, ever-growing human populations have interfered with its nesting grounds. (Koh Ngam Yai, Chumphon, in 12 meters.)

The banded coral snake (Laticauda colubrina), *like all other marine reptiles, evolved from land animals—creatures which, earlier still in evolutionary history, had moved from the sea to the land. This explains why sea snakes have lungs. They need air; and, although they can stay submerged for an hour or more, they generally come to surface to breathe every 20 or 30 minutes. All sea snakes are poisonous, some of much more so than the king cobra. They probably evolved such potent venoms because, if their prey did not die almost immediately once bitten, it would either get snapped up by another predator or else manage to flee to some inaccessible crevice. (Koh Ngam Noi, Chumphon, in 11 meters.)*

BIBLIOGRAPHY

GENERAL WORKS ON THE NATURAL HISTORY OF CORAL REEFS

Barnes, R.S.K. and Hughes R.N. 1982. *An Introduction to Marine Ecology*. Oxford: Blackwell Scientific Publications.

Coral Reefs of the World (3 Vols.). 1988. United Nations Environment Programme and the International Union for Conservation of Nature and Natural Resources. Cambridge, U.K.. "Thailand", Vol.2, pp. 371-384.

Levinton, Jeffrey S. 1982. *Marine Ecology*. Englewood Cliffs:Prentice-Hall.

Lowe-McConnell, R.H. 1979. *Ecology of Tropical Fishes in Tropical Waters*. London: Edward Arnold.

Mann, K.H. 1982. *Ecology of Coastal Waters: A Systems Approach*. Oxford: Blackwell.

Reader's Digest Book of the Great Barrier Reef. 1984. Sydney: Reader's Digest.

Roessler, Carl. Und*erwater Wilderness: Life Around the Great Reefs*. New York: Chanticleer Press.

White, Alan T. 1987. *Philippine Coral Reefs: A Natural History Guide*. Quezon City: New Day Publishers.

WORKS ON THE DISTRIBUTION AND CONDITION OF THAI REEFS

Baker, Ilyas and Pramuk Kaeoniam. Editors. 1986. *Manual of Coastal Development Planning and Management for Thailand*. Bangkok: Thai Institute of Scientific and Technical Research.

Brown, B.E., Le Tissier, M.D.A., Scoffin, T.P., Tudhope, A.W. 1990. "Evaluation of the Environmental Impact of Dredging on Intertidal Coral Reefs at Ko Phuket, Thailand, Using Ecological and Physiological Parameters". *Marine Ecology Progress Series*. Vol. 65: 273-281.

Chansang, Dr.Hansa, *et al.* 1988. *Living Resource in Coastal Areas With Emphasis on Mangrove and Coral Reef Ecosystems: Subproject on Inventory and Monitoring on Coral Reefs, Seagrasses and Soft Bottom Communities in the Andaman Sea*. Unpublished. Phuket: Phuket Marine Biological Center.

Chansang, Dr.Hansa, Boonyanate, P., and Charuchinda, M. 1985. "Features of Fringing Reefs in Shallow Water Environments of Phuket Island, the Andaman Sea". *Proceedings of the Fifth International Coral Reef Congress*, Tahiti. Vol.6: 439-444.

Chansang, Dr.Hansa, Boonyanate, P., *et al.* 1986. "Infestation of A*canthaster Planci* in the Andaman Sea". *Proceedings of the Second International Symposium on Indo-Pacific Biology.* Guam.

Chansang, Hansa, Boonyanate, P., and Charuchinda, M. "Effect of Sedimentation from Coastal Mining on Coral Reefs on the Northwestern Coast of Phuket Island, Thailand." 1981. *Proceedings of the Fourth International Coral Reef Symposium*, Manila. Vol.1: 129-136.

Chansang, Hansa. 1988. "Coastal Tin Mining and Marine Pollution in Thailand". *Ambio.* Vol.13, No.3: 223- 228.

Ditlev, Hans. 1976. "Stony Corals (Coelenterata: Scleractinia) From the West Coast of Thailand". Phuket: *Phuket Marine Biological Center Research Bulletin* #13.

Ditlev, Hans. 1978. "Zonation of Corals (Scleractinia: Coelenterata) on Intertidal Reef Flats at Ko Phuket, Eastern Indian Ocean". *Marine Biology*, 47: 29-39.

Lemay, Michele H., and Chansang, Dr.Hansa. 1989. *Coral Reef Protection Strategy For Phuket and Surrounding Islands.* Bangkok: Thailand Coastal Resources Management Project, Office of the National Environment Board and the University of Rhode Island, United States Agency for International Development.

McManus, John W. "Coral Reefs of the ASEAN Region: Status and Development". 1988. *Ambio.* Vol.17, No.3: 189-193.

Phonsuwan, Niphon, and Chansang, Hansa. "Coral Reef Resources of the Tarutao National Park, Thailand". Paper presented at the Symposium on Coral Reef Management in Southeast Asia, Bogor, Indonesia, April 22-24

Ruyabhorn, P. and Phantumvanit, D. 1988. "Coastal and Marine Resources of Thailand — Emerging Issues Facing an Industrializing Country". *Ambio.* Vol. 17(3): 229-232.

Sudara, Dr. Suraphol, *et al.* 1989. *Living Resource in Coastal Areas With Emphasis on Mangrove and Coral Reef Ecosystems: Site Selection Survey Report, The Gulf of Thailand.* Unpublished. Bangkok: ASEAN-Australia Cooperative Program on Marine Science. Coral Research Team: Department of Marine Science, Chulalongkorn University and the Department of Aquatic Science, Srinakharinwirot University.

1989. Report for In-House Workshop on Strategies for Coral Reef Protection, Coastal Resources Management Project.

1991. Thailand Coastal Resources Management Project, together with the University of Rhode Island, the Department of Technical and Economic Cooperation, and the United States Agency for International Development. *A National Coral Reef Strategy for Thailand.* Vol.1: *Statement of Need.* Bangkok.

General Index

Index of Scientific Names